COLLECTED POEMS
1988

David Gascoyne

Oxford   New York

OXFORD UNIVERSITY PRESS

1988

Oxford University Press, Walton Street, Oxford OX2 6DP

Oxford New York Toronto
Delhi Bombay Calcutta Madras Karachi
Petaling Jaya Singapore Hong Kong Tokyo
Nairobi Dar es Salaam Cape Town
Melbourne Auckland

and associated companies in
Beirut Berlin Ibadan Nicosia

Oxford is a trade mark of Oxford University Press

Poems © David Gascoyne 1965, 1988
Introductory Notes © David Gascoyne 1988

Collected Poems first published by Oxford University Press
in association with André Deutsch in 1965 and
reprinted 1966, 1970, 1978, 1982, 1984
This revised, expanded edition, Collected Poems 1988,
first published as an Oxford University Press
paperback 1988

British Library Cataloguing in Publication Data
Gascoyne, David, 1916–
Collected poems 1988.
I. Title
821'.912   PR6013.A75
ISBN 0–19–281972–0

Library of Congress Cataloging in Publication Data
Gascoyne, David, 1916–
(Poems. Selections)
Collected poems 1988 / David Gascoyne.
p.   cm.
Rev. and enl. ed. of: Collected poems. 1965
I. Title.
PR6013.A75A6   1988
821'.912—dc19   87–28441
ISBN 0–19–281972–0 (pbk.)

Set by Wyvern Typesetting Ltd.
Printed in Great Britain by
J. W. Arrowsmith Ltd. Bristol

To Judy, my wife

CONTENTS

IV HÖLDERLIN'S MADNESS (1937–1938)

INTRODUCTORY NOTES

THE first poem of mine to be accepted for publication was entitled 'Transformation Scene', and appeared in the literary weekly *Everyman*. In 1932, while still a day-boy at a West End secondary school, I persuaded an obscure publishing firm in a Court off the Charing Cross Road to publish, under the title *Roman Balcony*, a collection of poems including 'Transformation Scene'. My mother (who never considered herself to be much of a judge of poetry) told me: 'You'll only regret it later.' Before long this proved to be true. For many years after the mid-'30s, I did not wish this early 'slim volume' ever to be alluded to. During recent decades, however, *Roman Balcony* has from time to time appeared as a rarity in bookdealers' catalogues at ever more extravagantly high prices; which has encouraged me to reprint in the present collection nine of the forty or so items it comprises, though my choice includes neither 'Transformation Scene', 'Prison' (reprinted in Robin Skelton's Introduction to my *Collected Poems*, OUP 1965), nor 'Mood', republished recently in Jon Stallworthy's anthology *First Lines* (Carcanet 1987).

In April 1933, a weekly column called 'Poets' Corner', run by Victor B. Neuburg (an endearing eccentric once involved with the black-magician poet Aleister Crowley), began to appear in *The Sunday Referee*. Among the younger poets whose work subsequently appeared in this column were Dylan Thomas, Pamela Hansford Johnson, Ruthven Todd, Julian Symons, Laurie Lee, and myself. 'Slate', the first poem in *Collected Poems* of 1965, was first printed in *The Sunday Referee*. I now reprint for the first time 'Seaside Souvenir' and 'On the Terrace' (Richmond Terrace, near where I then lived), both of which also first appeared on a 1933 Sunday.

One of my mother's best friends had lodgings for many years when I was a boy in the house of Alida Monro, who with her husband Harold ran the Poetry Bookshop in Bloomsbury. I was first introduced to this shop at an early age, and once heard T. S. Eliot give a reading there, not of his own poetry but of Christina Rossetti's. In 1933 Mrs Monro, then not long a widow, edited *Recent Poetry: 1923–1933*, in which she magnanimously included three poems of mine, one of them 'Slate'. Barely seventeen, I must have been the youngest contributor to this anthology, arranged in alphabetical order and intended to represent a sequel to the *Georgian Poetry* series of Edward Marsh, and was undoubtedly gratified to find myself in the company of Yeats and Eliot, as well as of the poets of Auden's generation, and of George Barker, whose *30*

Preliminary Poems had just been published by David Archer's Parton Press, which three years later was to publish a small collection of my own.

1933 was something of an *annus mirabilis* for me. It was the year when Geoffrey Grigson, at that time working for *The Morning Post*, began publishing from his Keats Grove home his small, adventurous, and soon influential periodical *New Verse*. In one of its earliest issues Grigson published 'And the Seventh Dream is the Dream of Isis', the result of my first attempt to produce a sequence of lines of poetry according to the orthodox surrealist formula: 'Pure psychic automatism by which is intended to express . . . in writing . . . the real process of thought . . . in the absence of all control exercised by the reason and outside all moral or aesthetic preoccupations', in the words of André Breton, instigator of the surrealist movement. I was not to become a fully-fledged and committed member of this movement until two years later; but already before leaving school earlier that year I had been in the habit of visiting Zwemmer's bookshop in the Charing Cross Road, on my way home via Waterloo, to purchase not only back numbers of Eugene Jolas's avant-garde *transition* but also previous issues of *La Révolution surréaliste* (1924–9) and then of the more recent *Le Surréalisme au service de la Révolution*. In November 1933, A. R. Orage published in his *New English Weekly*, to which I was to become for a few years an occasional contributor, the series of short surrealist texts that in the present volume I have retitled 'Automatic Album Leaves'.

The semi-autobiographical stream-of-consciousness account of a day in the life of an adolescent literary aspirant to be found in *Opening Day*, my only novel, completed the year I left school, gives no indication of an awareness of Surrealism, though it contains a passage of enthusiastic reference to Rimbaud. After finishing it I had given it to Alida Monro to read, and she eventually decided to submit it to Cobden-Sanderson, who had just become the publishers of Harold Monro's posthumous *Collected Poems*, prefaced by Eliot. The novel was duly accepted; and the advance royalties I received on its publication from Cobden-Sanderson contributed to financing my first visit to Paris, where I was able to spend the last three months of 1933.

At this point I am tempted to digress into a detailed account of what was for me a momentous first encounter with France, a country in which I was subsequently to spend, on and off, at least fifteen years of my life. For the purpose of introducing the poems in the present collection, however, I must restrict myself to recording that although I did not then make initial personal contact with any of the representative writers of the surrealist group, I did visit Max Ernst's rue des Plantes studio for the first time, and brought away with me from it one of his gouaches, a *Oiseau en forêt*; and from a visit to the shop at the foot of Montmartre of the official surrealist bookseller and publisher, José Corti, I brought away copies of recent collections by such poets as Breton, Eluard, and Tzara.

Some of the poems arranged together for the first time in the present volume under the general heading 'Surrealist' were first collected (confusingly accompanied by a certain number of non-surrealist items) in the little book published under the title *Man's Life is This Meat* in the summer of 1936 at the time of the London International Surrealist Exhibition. All these poems are united by the basic aim of achieving the greatest possible spontaneity, but this aim can produce results of considerable variety. In 1935, Geoffrey Grigson published in *New Verse* 15 a group of short pieces of a type quite dissimilar from the apparently incoherent pellmell outpouring of images and phrases characteristic of 'And the Seventh Dream . . .'. Each of them appears to have some underlying theme or subject, though never a preconceived one. The title was usually added after the poem's completion, as is said to have been the case with the poetic pictures of Paul Klee. 'Gnu Opaque', for instance, was the watermark faintly distinguishable in the paper on which it was written. The title of the 1936 Parton Press collection was the result of a meeting with Geoffrey Grigson during which he produced a sample-book of printers' type-faces, which when opened at random showed the words 'man's life is' in one sort of type at the end of the bottom line on the left-hand page, and 'this meat' in a different type of lettering at the beginning of the top line of the page opposite: as an example of what the surrealists described as 'objective hazard', this seemed at the time an ideal title. 'The Truth is Blind' is a title applied without reflection to the result of an attempt to create a poem by adopting the technique of collage: three cuttings were selected at random from *Argosy Magazine*, *The Listener*, and an evening newspaper, which happened to be the sources nearest to hand at the time, and then stuck on two sheets of paper with spaces left between them to be filled in such a way as to link them into a more or less coherent whole, while avoiding stopping to consider anything like a normally logical connection between the three disparate component elements. A scarcely avoidable presupposition in this case was that the result would read like the account of a specific dream.

A French professor of English once asked me: Why did you call one of your poems by the name of a village near where I live outside Lyons? He was referring to 'Lozanne', the result of a specific conscious premeditation, to elucidate which requires some explanatory gloss. In 1933 there occurred in France a *cause célèbre* that, while scandalizing the general public, so aroused the indignation and sympathy of the surrealist group that its members collaborated in producing a collection of poems and drawings inspired by it. This was the case of Violette Nozière, put on trial for parricide and sentenced to life imprisonment. When in 1977 Louis Malle made a film based on the Nozière affair, portraying it as a classic instance of the triumphant hypocrisy of bourgeois morality, he made a point of referring in it to the poem contributed by Paul Eluard to the Surrealists' collective *plaquette* of protest and homage to the accused. In the summer of 1935, England was for weeks shocked and electrified by the Rattenbury/Stoner case (to be dramatized by Terrence Rattigan in his last play, *Cause Célèbre*). Readers of any British newspaper at the time would have been aware that Mrs Rattenbury had already made a name for herself as a composer of light music under the pseudonym 'Lozanne'; they would also have seen photos of her invariably wearing a slim bandeau across her forehead. After a trial resulting in her acquittal and the sentencing to death of her young chauffeur lover, Alma Rattenbury committed suicide by drowning. Though genuinely touched by her fate, I doubt whether I should have written 'Lozanne' had I not recently seen a copy of the Surrealists' *Violette Nozière*.

As it is no longer possible to present the poems in the Surrealist section of this book in strictly chronological order, I have found it preferable to place together the four poems inspired by or dedicated to painters. 'Charity Week' is inspired by the sequence of Ernst's collages entitled *Une Semaine de Bonté*, the 'hero' of which is the Lion of the Place de Belfort. 'Yves Tanguy' attempts to evoke the atmosphere of his earlier unearthly landscapes. 'Salvador Dali' was originally entitled 'In Defence of Humanism'; it does not attempt to present in verbal terms the imagery to be found in Dali's best-known works, but to provide some sort of parallel equivalent of the personal 'mythology' his paintings embody. To each of the six stanzas of 'The Very Image' the title of one of Magritte's pictures could be affixed, though I had no idea when starting the poem what images were going to occur to me in the course of writing it: I had decided in advance only that each stanza should have five lines.

A similarly convenient grouping together is that of all the items in prose. I hesitate to designate them 'prose poems', since this category has been denounced cogently and with wit by George Barker as representing a 'Jubjub Bird'. The sequence now retitled 'Automatic Album Leaves' is no more than early exercises in uncontrolled word-play. 'Reflected Vehemence' probably represents the most successful of my attempts to register what Breton called *'le fonctionnement réel de la pensée'*; it was written in haste, without hesitation or the least intention to mystify, though its content defies analysis. The longest piece, 'The Great Day', was obviously written in emulation of the texts in *L'imma-culée conception*, produced in collaboration by Breton and Eluard with the intention of simulating various types of mental disorder. Paranoia would appear to be the most easily imitable of such derangements. The pieces retitled 'Three Verbal Objects' were first published in the catalogue to an Exhibition of Surrealist Objects at the London Gallery in the winter of 1937, by which time I had moved to a Paris attic and virtually ceased writing in the surrealist vein. They are posthumously dedicated to Humphrey Jennings, in acknowledgement of the influence that his 'Reports' and other admirable short texts, first published by Roger Roughton in *Contemporary Poetry and Prose*, undoubtedly had on me.

'The Supposed Being' first appeared in the original *Faber Book of Modern Verse*, edited by Michael Roberts and published in 1935. 'The Symptomatic World' was originally planned as a sequence of a dozen parts, each to be written at a session. Some appeared in the short-lived review *Janus*, some in Roger Roughton's magazine; the remainder appear to have been lost. 'Phantasmagoria' was written early in 1939, when I had returned from Paris and was no longer writing poetry classifiable as surrealist. A young friend of friends insisted that I should write a poem especially for her. Unable to produce a suitable poem to order, I proceeded to employ the formula of quasi-automatism I had been accustomed to use during four previous years. The deliberate repetition of such a motif as a little black town on the edge of the sea is a device I would not formerly have allowed myself (except perhaps in the poem about sleep, 'Unspoken'). John Lehmann included it in the 1942 issue of *Poets of Tomorrow*, together with other poems of a quite different description.

In the autumn of 1937, my discovery of a copy of the 1930 edition of Pierre Jean Jouve's *Poèmes de la Folie de Hölderlin* in a book-dealer's box on the Paris quays marked a turning-point in my approach to poetry. I had not so much become disillusioned with Surrealism as begun to wish to explore other territories than the sub- or unconscious, the oneiric and the aleatory. Jouve's Hölderlin translations led not only to my essay, poems, and translations published by Dent the following year as *Hölderlin's Madness*, but to an excited first reading of Jouve's own poetry and prose, and before long to an acquaintance with the poet and his psychiatrist wife that was to last nearly thirty years. The use of lines quoted from Jouve as epigraphs to certain sections of *Poems 1937–1942* is insufficient indication of the enormous influence that his poetry, outlook, and conversation were to have on me for many years to come. Anyone familiar with Jouve's *Sueur de Sang, Matière Céleste* or *Kyrie* will recognize this influence in such poems of mine as 'World Without End', 'The Fortress', and 'Insurrection'.

This is not the place to pay further homage to a poet I still regard as the greatest it has been my good fortune to know. I should however add that 'The Fabulous Glass' now appears, as it should have done from the first, with a dedication to his wife, Dr Blanche Reverchon, as it represents a half-rhyming versification of a sequence of images that actually occurred to me during a psychoanalytic session with her in late

1938 and noted down immediately after: the Virgin and child in an alcove were in fact a medieval statuette of the Virgin with her child's face obliterated by an iconoclast or time, treasured by Jouve and kept in a recess in the study adjacent to his wife's consulting-room, to become the inspiration of his collection *La Vierge de Paris* (1939–44).

The place of 'The Conspirators' in the present collection should strictly speaking be between 'Snow in Europe' and 'Farewell Chorus'. I first read W. H. Auden's paperbacked *Poems* and his *The Orators* soon after their appearance, the *New Signatures* and *New Country* anthologies likewise; and it is not improbable that, had I not been carried away by enthusiasm for contemporary French poetry, and for *le surréalisme* in particular, I should have endeavoured to find a way of my own to express the politico-social awareness cultivated by many of my contemporaries and their immediate predecessors. I was as keenly conscious as they were of the meaning of current events in Europe, as well as of the hunger marches, and the menace of Mosley at home. In the summer immediately preceding the outbreak of war with Germany, in my family's home at Teddington, I was seized by an anomalous impulse to embark on a long narrative poem to be entitled 'Come Dungeon Dark'. Its setting was an imaginary European country on the brink of a fascist coup and the installation of a reign of terror and tyranny. The hero was to be a left-wing social scientist with the impossibly romantic name of Flambow. If my memory still serves me faithfully, this character, after the take-over of his country by dictatorship, was to retreat with a band of his comrades into hiding in a disused mine, finally to emerge, after numerous Resistance-type sorties and forays followed by a disastrous flood, to inaugurate the triumph of socialism after the dictator's downfall. That I could ever have carried out such a scenario in verse was of course a delusion; but I reprint the pages I did succeed in writing because they not only convey something of the atmosphere peculiar to the period, but also represent a reminder of my brief involvement with Mass Observation during those evenings in Blackheath in late 1936 when Charles Madge and Humphrey Jennings were about to launch it as a movement (in which later I took little part). The introductory episode of this unfinished epic was published by John Lehmann in the new series of his *New Writing* in the winter of 1939, and remained unreprinted and forgotten until recently.

A couple of poems that remained similarly forgotten until recently are 'Elsewhere' and 'Concert of Angels'. They appeared originally in Miron Grindea's *Adam International Review* just after the end of the War, though they may have been written earlier, perhaps at about the same time as the 'Requiem' later set to music by Priaulx Rainier. The second is recognizable as having been inspired by one of the panels of Grünewald's Isenheim altarpiece. The 'horrifying face, discoloured, flayed', in 'Ecce Homo' was likewise the result of having been impressed by the central figure in the black-and-white reproductions of this masterpiece that I first saw before the War. It was not until at least ten years later that I was taken to Colmar to see the original. 'Elsewhere' is an unmitigated overstatement of an underlying theme that has remained constant in almost everything I have written since then: the intolerable nature of human reality when devoid of all spiritual, metaphysical dimension.

I first returned to Paris after the War in 1947, and remained there for a year. At least half-a-dozen of the items collected in 1950 as *A Vagrant & other poems* were written during this visit or as a result of it. 'A Vagrant' represents the apologia of a premature beatnik or drop-out, and is partly based on the idle, hotel-room existence I led at that time, increasingly disappointed with post-war governments' failure to implement the dreams and promises of a radically improved new future that had helped the Allies bring the Third Reich to an end. In my case this disappointment was compounded by the realization that I could no longer depend on the untramelled spontaneity of inspiration I had assiduously cultivated before the War. During the War, Tambimuttu's Poetry London Editions had published my *Poems 1937–42*, illustrated by Graham Sutherland; after which I had little time to write poetry as, unfit for military service, I turned professional actor for a couple of years, adopting Emery as my stage-name after that of my mother's family. The return of genuinely gifted demobilized young actors after the War meant that I was soon once more out of regular employment. I mention this in passing only because my intention during this period was to prepare myself through first-hand experience to contribute something to the revival of poetic drama that was still in the air at that time. The only result of this ambition was the production in 1950 of a satirical one-act piece concerned with the state of English theatre just before the abolition of censorship and the renaissance brought about by John Osborne and his contemporaries and successors. All that remains of *The Hole in the Fourth Wall*, as this production was called, is one of the Cabaret songs to be found here under the heading Light Verse.

The city setting of 'A Vagrant' is identifiable as Paris from its reference to straying 'slowly along the quais towards the ends of afternoons'. Young visitors to present-day France may no longer come across the 'cosy-corner', a franglais expression applied to a combined bed-head and book-shelves once a familiar feature of Parisian hotel rooms and bed-sits, and so fail to understand the allusion to a 'cosy-corner crow's-nest' also occurring towards the end of the poem. One or two other poems of this period require slight elucidation. The intention behind 'Innocence and Experience' was to produce something in the tradition of Eliot's early 'Portrait of a Lady', modelled on my experience of a couple of meetings with a certain Mme X, the wife of the owner-director of one of the best-known Parisian department stores. The line 'I still knew of her nothing less than this' leads to a complicated image intended to suggest a combination of two well-known portraits, one of Ellen Terry in the role of Portia, the other of the cellist Suggia by Augustus John, each of them intimating an aspect of Mme X's character and appearance. The imagined incident from her childhood is purely speculative. The setting is the *hôtel particulier* in the Faubourg Saint-Germain district in which she had lived for many years. The occasion narrated is close to what actually occurred when another lady belonging to her circle took me with her to call on Mme X again, for the first time in ten years. The works of art referred to were almost exactly as described.

The following piece, 'Photograph', was inspired by a portrait of Philippe Soupault in his prime by the American photographer Berenice Abbott. When I wrote this poem, which deliberately avoids anything visually concrete except the subject's eyes, I had still never met Soupault, who at the time of my frequenting the surrealist group in 1935/6 had fallen out of its official favour, having begun to write travel journalism and novels that might have been intended to be commercially successful. When I was finally able to visit him, he was in his mid-eighties. He will be ninety-one this year, the last survivor of the original surrealist movement, except Dali. A final explanation here: 'The Other Larry' refers to Lawrence Durrell. How I can have expected the reader to realize this in the absence of footnote or formal dedication, I don't know. The poem is in a sense an answer to one by Durrell, dated 1939 and addressed to me (his 'Paris Journal': *Collected Poems*, Faber & Faber 1957), and it attempts to sum up certain differences between our points of view that had first become apparent during our discussions in pre-War Paris. It is republished in the hope that its argument is of sufficient interest to be appreciated without consideration of the specific persons involved in it.

In the autumn of 1951, I accompanied Kathleen Raine and W. S. Graham to America, where we gave a series of readings in New York and certain NE States under the guise of 'Three Younger British Poets'. I returned to England a year later, having first gone on from the States to Vancouver Island BC to visit my parents who at that time were living in retirement there. It was immediately after arriving back in this country that I learnt of the death of Paul Eluard. I had met him again only once since the War, during which his fame and popularity had increased enormously. So had his commitment to the PCF and Stalinism, largely, it seemed to me, as a result of his third marriage. The 'Elegiac Improvisation' I wrote after his death was an expression of the admiration of him I had first felt when not yet twenty. Passages of the poem use brief lines imitative of his *Poésie ininterrompue*; others introduce imagery derived from the kind of French painting he loved and interpreted so well. It refers to him as the great poet of the Resistance that he was commonly supposed to be. It was not until quite recently, on reading Milan Kundera's *Book of Laughter and Forgetting*, which contains a bitterly ironic account of Eluard's inexcusable failure to speak out in defence of his one-time friend the surrealist Zavis Kalandra, who was hanged in Prague in 1950 during the French poet's visit to the city at the invitation of the Czech authorities, that I fully realized what kind of man he had become at the end of his life. If I had been aware of this incident at the time, and fully understood the way authoritarian politics can transform even so fraternal a poet as Eluard, it would not have been possible (or, at least, I hope not) for me to write the kind of poem that the 'Elegiac Improvisation' turned out to be. The poem was intended for recital and I first read it at the Institute of Contemporary Art, then still located in Mayfair; it was later published in the review *Botteghe Oscure*.

Soon after my return from America, Douglas Cleverdon of the BBC commissioned me to write a work for voices and music for the Third Programme. This turned out to be 'Night Thoughts', written in a relatively short space of time, and with the exception of the Eluard elegy the only poem of any kind I had been able to write since 1950. It was finally broadcast in December 1955, with music specially composed by Humphrey Searle. By that time I had gone to live in France, and was to spend the summer in Aix-en-Provence, the winter in Paris, for ten consecutive years, except for occasional brief return visits to England. During this period I was incapable of writing a line due to the block, or *crampe* as the French call it, that had resulted from a long abuse of amphetamines dating from as far back as the beginning of the War. In one of the 50-odd 'aphorisms' collected in *The Sun at Midnight* (edition limited to 350 copies, Enitharmon Press 1970), I discuss this addiction at some length, explaining that amphetamines 'have powerful and most undesirable side-effects which probably were responsible for reducing my output to the strict minimum of work on which a poet's reputation can plausibly rest'.

The title of the fragment entitled 'Half-an-Hour', dedicated to my generous hostess during those years of unproductivity, derives from one of the most mysterious phrases to be found in the Book of Revelation: 'And there was silence in heaven for the space of half-an-hour.' It is all that remains of an attempt to break my silence of years by exploring its nature and conditions. 'Remembering the Dead' was my only contribution to David Wright's review *X*, in which it appeared in 1959. The poem first published in 1970 in *Penguin Modern Poets 17*, under the optimistic title 'Part of a Poem in Progress', now changed to 'Unfinished Poem from Elsewhere', had suddenly emerged as though by dictation from the unconscious, unexpected and inexplicable, in 1964, just before the onset of a severe nervous breakdown, as a result of which I had to return permanently to England.

The 'Three Verbal Sonatinas' (which conclude the Light Verse section) were written in 1969 when I was convalescing in a psychiatric hospital from a further breakdown as chronic as that of five years previously. Several of the small number of poems produced since my final recovery from a third breakdown and my marriage in 1975 were written as a result of requests from editors. 'Whales and Dolphins' was produced for the Greenpeace organization's enormous anthology *Whales A Celebration* (Hutchinson 1983). The tribute to Miron Grindea was composed on the occasion of his seventy-fifth birthday, and printed on the cover of a special 45th anniversary number of his *Adam International Review* (also 1983). Its title is intended to indicate that it is what could be called a 'verbal square', consisting of twelve dodecasyllabic lines, or alexandrines, a form I had used the previous year to contain a comment on the Falklands conflict. The latter appeared at the end of a contribution to a compilation called *Authors take Sides on the Falklands*. 'A Sarum Sestina' was written specially for Satish Kumar's anthology *Learning by Heart*, published in 1984 to raise funds for The Small School founded by him in Hartland, near Bideford in Devon. Similarly, 'Thalassa: The Unspeakable Sea' was written simultaneously in English and French for the international anthology *Thalatta (Hommage à la Mer)*, published by Editions Internationales Eureditor of Luxemburg on the occasion of the 8th Congress of the World Organization of Poets held in Corfu in 1985, though it did not reach the editor in time to be included in it; the French version (*'Au delà de toute expression'*) eventually appeared later that year in no. 35 of the review *Phréatique*, and the English in *Temenos 7*. It is dedicated to Mimmo Morina, Secretary General of the World Organization of Poets. Finally, 'Entrance to a Lane' resulted from a request for a contribution to the anthology *With a Poet's Eye* (Tate Gallery, 1986).

The seventh *verset* of 'Thalassa: The Unspeakable Sea' combines allusions to Prospero's book of magic spells, to two of the Fragments of Heraclitus, and to Tennyson's early poem 'The Kraken'. 'A Further Frontier' was inspired by the view of the frontier to be seen from the North of the isle of Corfu of the frontier dividing mainland Greece from Albania. Greek-hay is a variant of fenugreek, a herbal plant the green of which is distinguishable from that of coniferous foliage. The last lines of the poem derive from the conclusion of Schiller's lyric *Gruppe aus dem Tartarus*, set as a *lied* by Schubert. 'November in Devon' contains a reference to an autumn landscape clad in the colours of DPM, the military term for 'disruptive pattern material', in other words the camouflage-type stuff of uniforms now worn by troops and guerrillas throughout the entire civilized world.

In addition to those I have already acknowledged throughout these introductory notes, I would like to thank particularly my bibliographer Colin Benford, Alan Clodd of the Enitharmon Press, and Professor Norma Rinsler of King's College, London; and belatedly, Robin Skelton, whose edition of the first *Collected Poems* has paved the way for this fuller and more complete edition of my poems from 1932 to 1986.

DAVID GASCOYNE,
Isle of Wight, July 1987

I
ROMAN BALCONY
(1932)

FIVE NETSUKÉS OF HOTTARA SONJA

1

No larger than my finger-nail
this little face so finely carved,
more pink than the cherry in bloom.

Finer than silk the tiny hairs
smoothly over the forehead combed,
as delicate as spiders' thread.

What lovely landscapes do you see,
O tiny gazing Eastern eyes,
so constant in your ivory?

2

They cut you in halves, O monstrous pumpkin,
the strong little men in cotton jackets,
they cut you in halves with their cruel saw.

The jade-green leaves, so small and fragile,
cling to the sides of their dear old pumpkin,
they fear lest the saw should split him in two.

A black butterfly on one side flutters,
on the other side a white butterfly,
endeavouring to hold him together.

3

Do not fear, little ape,
do not be so timid.
Those five pointing pillars
are fingers of a hand.

Do not fear, little ape,
this mysterious thing.
It is a human hand,
but its owner is dead.

Do not fear, little ape,
so neatly dressed in silk.
With the long stalk prod it
that you hold in your paw.

4

Greatly do fishermen fear the mermaid.
They say she is the sorceress of the sea.
She has long, untidy hair, like seaweed.
Many are the silver scales which cover her body.
Her tail lies curled across her back.
She lies on her stomach, resting her chin on her hands,
staring at us with cunning, curious eyes.

5

With her small hands, an elegant lady
sits making music on her samisen,
smiling at the melodies awakened from the strings.

Wryly tasting the half of a lemon
her playful tame monkey sits at her side
dressed in a tunic embroidered with red strawberries.

But Death, a skeleton carefully carved,
leans over, leering, from behind, unseen,
hiding the grin of his teeth with an ivory fan.

FADING AVENUES

At my feet, trembling in the wind,
lies a rusty and serrated leaf,
alive with sun-caught moisture,
with a scarlet stem.
Above my head as I stand, cold, dreaming,
a tattered projection of black-spotted leaves
on a branch.

The avenues are fading
and my sight is fading fast as they
for I see but vaguely the figures that pass:
... There is a crimson coat ...

The sound of the wind is like water; ...
(water falling only in dreams,
for the fountain is choked, the fountain is stained,
at its foot a few burrs rotting lie).
The sound of the wind is fading,
and fading the sad sound of feet
drifting over the lawns
where grey's on the sheen of green.
The sound of the wind is fading ...

The wind creeps slowly up my spine
and creeps up the boles of the trees.
The trees stand brooding over their disintegration:

The ichor within grows lifeless and cold.
Above them one pine exulting stands
for the green of its foliage never fades.
But the avenues are fading
and the mould of the flower-beds is sour and dark
and the stems of the shrubs are black
with a sudden ignition of leaves at their tips.
The avenues are fading and sounds drift from afar.

Whose tomb shall we discover
in the dun shade of the woods
at the end of the fading avenues?

SUMMER'S ECHO

Cold is the day,
colder than fires of water,
colder than ashes of a forgotten moon.
In the dark room under the tower
the shutters flap in the draught.
The hollyhocks hang broken

Empty,
void in space a sound (trickling
through colossal stretches
of arid air) intimating
some tremendous music
beyond our consciousness.

Some white figure with long hair
walks through the mist
sighing and stirring the branches
of sleep, walks through the room
raising the dust from the stones
of the cold-paved floor.

BEFORE STORM

Wet wind
    dark clouds trembling on darker
    horizons where the great trees
    sigh in the forests and shiver
    and let fall their leaves
    into swollen streams.

Cold sword
    stabbing the bare white light
    churning the almost ice.

Wet wind gathering its sombre draperies above
the plain where the thin rain has begun to
scatter its silver drop by drop.

    Whither wet wind? . . .

EVENING ON THE THAMES

Mud-flats mirroring
the blue reflecting
the sheen of the
evening waters
where the ripples
of our boat course
shining and collapse
The slender cranes . . .
black, the slanting
masts, the distant
dome, the tiny
figures on the
barges at rest
The solemn rays
that descend
beyond the misty bridge . . .
the waters and the wharfs . . .
                    the waters . . .

ECLIPSE OF THE MOON

Tonight the moon
will hide her face
behind a veil
of cloudy lace.

Blow out that candle;
let blue darkness swim,
silent as water
to the window's rim
where a blurred ray
of moonlight falls
making of glass
enchanted walls
which hold its radiance
yet strangely divide
this inner darkness
from the silver outside.

Beyond the window
the gardens are still
as the misty waters
which dewponds fill.

The trees have hid
their moonlit green
behind a film
of greyish sheen.

The lake has ceased
to echo back
the heavens' chequered
white and black.
The rose that shone
whiter than snow
has lost its strange
and inner glow.

Beyond the trees
the moon grows dim;
a shadow creeps
across its rim.
It has begun,
the moon's eclipse:
She lifts to her face
her finger tips.
Soon she'll illumine
the empty night
no more than my
blown-out candlelight.

REFLECTED IN JET

Draw back this curtain:
Here the houris languish,
watching their reflections
in amber or in jet.
'Ours for ever
are myrtle and scarlet,
and linked chains of passion-flowers,
and madrigals or mandolins.'

On our side of the curtain
we contemplate carefully
one tall glass of wine . . .
Names, names, names,

    Nemesis.

PLETHORA

Your hand in the mirror
plain and clear
five fingers and a palm

      *

not any pillar of smoke
will guide in this
the wilderness of your room

      *

where five fingers sharply
cut the air
between reflection and the
fire's rising dust.

      *

LUCUBRATION

The fir-branch projecting,
attendant, protecting;
the old grief releasing
stale memories, increasing
old ennui and spleen;
the fir-branch's green
and the butterfly's white;—
my words in the night.

The chair by the fireplace;
the perpetual clock-face;
my prison:—this room
where a fir-branch's gloom
and a butterfly's wing
induce me to sing,
induce me to write
these words in the night.

II
EARLY POEMS
(1933–1938)

SEASIDE SOUVENIR

The pattern the jelly-fish left behind,
a pocketful of sand,
a dead, pressed leaf,
the woven rhythms of three days:
these are their traces, faded, indistinct.

The cliff's wide boulders, the immense
rocking of ocean through the bay,
the lighthouse beam stabbing the rainy night:
these are the memories of three days and more,
not separate, but one—and quite distinct.

ON THE TERRACE

A heavy day: so old the sky
That covers up the treegrown leagues below;
So cold the figures up and down
The terrace where the gusty fountains blow.

Here comes a colonel, at his side
His wife, with drooping shoulders, dressed in black.
They neither of them speak a word.
The colonel walks with hands behind his back.

The woman wears a fading rose
Upon her breast. She and the colonel stare,
Dumb, at the footworn pavingstones
As they walk on. A sigh disturbs the air.

Stirred by no dull regrets for youth,
Or love now dead that once in Spring was new,
Too tired to speak of memories,
They pause and turn to contemplate the view.

Then they pass on. A fountain leans
To drench the stones on which they stood with spray;
And from an ironrailinged tree
A bird looks after them—then flaps away.

SLATE

Behind the higher hill
sky slides away to fringe of crumbling cloud;
out of the gorse-grown slope
the quarry bites its tessellated tiers.

The rain-eroded slate packs loose and flat
in broken sheets and frigid swathes of stone,
like withered petals of a great grey flower.

The quarry is deserted now; within
a scooped-out niche of rubble, dust and silt
a single slate-roofed hut to ruin falls.

A petrified chaos
the quarry is; the slate makes still-born waves,
or crumbling clouds like those
behind the hill, monotonously grey.

IN PERPETUUM MOBILE

Too tightly tangled are mixed motions;
Wide ocean's wrack-worn tracks trace whorling wheels;
The vampire sun sucks up the sea's salt scum
And twists it into cloud that rolls or reels
In woven webs across the crystal sky;
The sun's barbaric cockerel comb of fire
Royally rages, reaching myriad miles,
Revolving regent rays that outwardly expire;
The system which has sun for centre spins
Round other systems which are cogs for more
Which act on others to the orbit's end,—
Continual correlation, endless war.
Unending Motion changes as it goes,
Like glyptic flame or shifting waterfall;
One moment is, then metamorphosis
Alters what was before to not at all.
Disintegration is the uncertain seed
Of Motion, making all seen things seem
A nystagmus, leaving no proof to show
That what we saw or shall see is not dream.

THE COLD RENUNCIATORY BEAUTY

The cold renunciatory beauty of those who would die
to hide their love from scornful fingers of the drab
is not that which glistens like wing or leaf in eyes
of erotic statues standing breast to chest
on high and open mountainside.

Complex draws tighter like a steel wire mesh
about the awkward bodies of those born under shame,
striping the tender flesh with blood like tears
flowing; their love they dare not name;
Each is divided by desire and fear.

The young sons of the hopeless blind shall strike
matches in the marble corridor and find
their bodies cool and white as the stone walls,
and shall embrace, emerging like mingled springs
on to the height to face the fearless sun.

LIGHT OF THE SUN OVER ARCTIC REGIONS

Light of the sun over arctic regions
Presides, striking the sides of icebergs
With slanting oblique rays, setting
The opaque snow translucently aglow,
Illumining blocks sedate in indigo depths.

There the unending fields of frost are blown
Upon by the harsh desolate blast;
The sun lacks warmth; alone at last
With wind from beyond, night from above and below,
Snow's light is negative, white equals black.

On the heart's bitter winter shines love's face.
Breaking, a berg groans response;
A facet's radiance, a moment's melting
Are answer. Soon gone is the sun.
The frigid heart feels death's wind only.

MORNING DISSERTATION

Wakening, peering through eye-windows, uncurious, not amazed,
Balance the day, know you lie there, think: I'm on earth.
Remember death walks in the daylight, and life still through filter
  seeps,
While you will remain unchanged, perhaps, throughout the day.
Time like an urgent finger moves across the chart,
But you are you, Time is not yours alone,
You are but one dot on the complex diagram.

Then are you a star, a nucleus, centre of moving points?
Are you a rock-crumb, broken from cliff, alone?
Or are you the point of a greater star, moving in unison?
If you are isolate, only a self, then petrify there where you stand;
Destinies crumble and bodies run down, the single sconces burn
  out,
But you are complete if without you completion is lacking,
Then you burn with the perfect light and are Time's bodyman.

SPECULATION

By marking off this footstep from that
And various other efficiencies of the day
One can easily dismiss the mind's insistence
As to direction, one can protect
The suspicious eye by diversion from the
Horizon's symmetrical doom, its carefully
Draped clouds, patterned stain, the rain
Coming in curtains of downward arrows, not
Yet felt upon the skin.

                              Though one must hear
Distant thunder, when the alert attention
Is drawn away from its visible manifesto
That can mean not omen, not cannon.

Can mean bold perhaps music or heavy
Traffic of increased commerce or crass
Stupid bodies' collapse of those we loathed.
Shortly the arriving rain will lay a chill
Finger on the unprepared skin, then up-
ward focused eye, annoyed at disturbance will
Appreciate storm's reality, appreciating
Folly past not fully, thinking Here's a
Splendid show, what grandeur Nature in this mood
Displays! gazing around in idiot wonder till
The sudden lightning shatters skulls,
Melts bones, coagulates all blood.

A SUDDEN SQUALL

After some days of heat
Withering leaf and bloom
Like pebbles falls the hail,
Like chips of stone the sleet
Out of the sudden gloom
Across the peaceful vale
Just now so bright.

While we are waiting for
The sulky storm to stop
Hour after hour,
Watching the garden lake
Toss the toy ship,
The orchard fast falls dark
And bruised fruits drop.

Birds are all flown;
Rabbits in holes
Wait for the sun's return;
At sea great whales
Send up their fountains
As they drive taciturn
Through waves like mountains.

Green becomes sodden grey
And across the fields
At death of day
Mist draws its chilly sheets,
And darkness wields
Its eerie power, night's
Creatures begin to cry.

This weather's change is blind.
His hopes grow dimmer
Who thought that summer
Might have no end;
Would have good reason
To resign his mind
To a rainy season.

LANDSCAPE

Across the correct perspective to the painted sky
Scores of reflected bridges merging
One into the other pass, and crowds with flags
Rush over them, and clouds like acrobats
Swing on an invisible trapeze.

The light like a sharpened pencil
Writes histories of darkness on the wall,
While walls fall inwards, septic wounds
Burst open like sewn mouths, and rain
Eternally descends through planetary space.

We ask: Whence comes this light?
Whence comes the rain, the planetary
Silences, these aqueous monograms
Of our unique and isolated selves?
Only a dusty statue lifts and drops its hand . . .

THE UNATTAINED

On the evening of a day on the threshold of Summer,
Before the full blast of vertiginous Summer, I flung
This foursquare body down upon the crumpled ground,
Moist with a dew-like sweat; and on all sides heard
The ceaseless clicking and fret of insect swarms;
I felt energy drain from these limbs spread cruciform,
Dribble away like sap from crushed bracken's veins;
Felt this my heaviness upon acid-green grass and sand,
Under the passive sky, becoming magnetic as stone;
And my lids slid down over eyes fanned by coloured winds.

And fierce desires swelled up from out my quiet:
To pierce through this flesh outwards, to embrace
The eternal blue, against my nostrils to smother
The fragrant cotton of the clouds; to feel beneath
Impatient soles of feet the grinding grit
Of gravel, the sharp sides of stones; and without end
Against the eyeballs' skin to press fresh images,
To lave in the swift stream of forms these avid eyes:
By passion suspended, hands stretched out, gnawed
From within, O how and to where could I pass?

Not within facile grasp swings that unattainable globe:
Though to catch an echo of the spheres' music these ears strain
And nostrils yearn for the rich scent of flame and of blood,
Hands strive clumsily phantom's ambiguous flesh to caress,
In vain the inward divinity batters against the gates,
Kicking against the pricks until the urgent spirit breaks.
Hourly the ocean, World's clock, smashes against the cliffs;
And savage relentless Time shreds onwards through the skull,
Whispers: 'Come home, only Death burns out there'. And I know
That this is my body, my cell, and I am alone and prone.

REINTEGRATION

After a plenitude of defeat, a load of sorrow.
Forget your coward victories, your crown of thorns,
And send the sulky eye-witness away;
Block out that solitary figure, the proud
Indomitable one. Hack down the heavy black
Statue. And because you can only remember
The darkest days of defeat, your weariness,
Because you can see but death's sinister finger
Always pointing to the shadowed wall,
Raise no more gloomy monuments, or build
A more transparent wall.
                 And listen
To the rich voice like flute-voice breaking
Suddenly from the white marble larynx;
Sunlight breaking suddenly upon the naked torso
Like the rustling down of a flimsy dress.
Listening, join proud singing with the voice,
As the sound of an inland sea now freed,
Smashing its winter cage of ice and rushing
With liquid arms and hands of foam uplifted
Across the frozen lands toward the outer seas.

THE HERO

The laurelled profile with the Caesar's nose and lip
Beneath the garlanded triumphal arch
Is not the Hero, for he has no face
But is as featureless as light.
Only the hands,
Stretched out before him in unending process of
Possessing all, are human as hands are: only
The hands, the heart
Which turns from side to side like searchlight rays,
Unresting, through the night, proclaim him man,
Because the man has died.

He is unknown in death. He brings
No music with him. But he seems
Still listening to the moment of the vast
Explosion which has snatched him out of life,
So hugely deafening that it cannot end
But is forever everywhere,
As the dust of a lost glory fills
Even the crevices of furthest stars.

SIGNS

There fell down on the shadowed sand
Like dead birds from an evil nest
Across a livid space of sky:
A writhing hand,
The pale globe of a breast.
And a dismembered thigh.

But from the dark's most secret place
Across the curtains of the air
There presently began to rise
A dream-transfigured face
With lips exhaling prayer
And lambent eyes.

III
SURREALIST POEMS
(1933–1936)

AND THE SEVENTH DREAM IS THE
DREAM OF ISIS

I

white curtains of infinite fatigue
dominating the starborn heritage of the colonies of St Francis
white curtains of tortured destinies
inheriting the calamities of the plagues of the desert
encourage the waistlines of women to expand
and the eyes of men to enlarge like pocket-cameras
teach children to sin at the age of five
to cut out the eyes of their sisters with nail-scissors
to run into the streets and offer themselves to unfrocked priests
teach insects to invade the deathbeds of rich spinsters
and to engrave the foreheads of their footmen with purple signs
for the year is open the year is complete
the year is full of unforeseen happenings
and the time of earthquakes is at hand

today is the day when the streets are full of hearses
and when women cover their ring fingers with pieces of silk
when the doors fall off their hinges in ruined cathedrals
when hosts of white birds fly across the ocean from america
and make their nests in the trees of public gardens
the pavements of cities are covered with needles
the reservoirs are full of human hair
fumes of sulphur envelop the houses of ill-fame
out of which bloodred lilies appear.

across the square where crowds are dying in thousands
a man is walking a tightrope covered with moths

2

there is an explosion of geraniums in the ballroom of the hotel
there is an extremely unpleasant odour of decaying meat
arising from the depetalled flower growing out of her ear
her arms are like pieces of sandpaper
or wings of leprous birds in taxis
and when she sings her hair stands on end
and lights itself with a million little lamps like glow-worms
you must always write the last two letters of her christian name
upside down with a blue pencil

she was standing at the window clothed only in a ribbon
she was burning the eyes of snails in a candle
she was eating the excrement of dogs and horses
she was writing a letter to the president of france

3

the edges of leaves must be examined through microscopes
in order to see the stains made by dying flies
at the other end of the tube is a woman bathing her husband
and a box of newspapers covered with handwriting
when an angel writes the word TOBACCO across the sky
the sea becomes covered with patches of dandruff
the trunks of trees bust open to release streams of milk
little girls stick photographs of genitals to the windows of their
     homes
prayerbooks in churches open themselves at the death service
and virgins cover their parents' beds with tealeaves
there is an extraordinary epidemic of tuberculosis in yorkshire
where medical dictionaries are banned from the public libraries
and salt turns a pale violet colour every day at seven o'clock
when the hearts of troubadours unfold like soaked mattresses
when the leaven of the gruesome slum-visitors
and the wings of private airplanes look like shoeleather
shoeleather on which pentagrams have been drawn
shoeleather covered with vomitings of hedgehogs
shoeleather used for decorating wedding-cakes
and the gums of queens like glass marbles
queens whose wrists are chained to the walls of houses
and whose fingernails are covered with little drawings of flowers
we rejoice to receive the blessing of criminals
and we illuminate the roofs of convents when they are hung
we look through a telescope on which the lord's prayer has been
     written
and we see an old woman making a scarecrow
on a mountain near a village in the middle of spain
we see an elephant killing a stag-beetle
by letting hot tears fall onto the small of its back
we see a large cocoa-tin full of shapeless lumps of wax
there is a horrible dentist walking out of a ship's funnel
and leaving behind him footsteps which make noises
on account of his accent he was discharged from the sanatorium
and sent to examine the methods of cannibals
so that wreaths of passion-flowers were floating in the darkness

giving terrible illnesses to the possessors of pistols
so that large quantities of rats disguised as pigeons
were sold to various customers from neighbouring towns
who were adepts at painting gothic letters on screens
and at tying up parcels with pieces of grass
we told them to cut off the buttons on their trousers
but they swore in our faces and took off their shoes
whereupon the whole place was stifled with vast clouds of smoke
and with theatres and eggshells and droppings of eagles
and the drums of the hospitals were broken like glass
and glass were the faces in the last looking-glass.

GERMINAL

In a manner of speaking
    I should in that manner indicate
    That which has processed through my skull
    Yesterday entering at the eyes and ears
    Issuing tomorrow from the mouth

    The marvellous is yet unborn
    In the Manor of the Tongue
    Seed fallen until now on stony ground

    Spoken then
    An announcement of future marvels.

GNU OPAQUE

No more resistance
    No letters this morning
    Tomorrow will be a fine day

    Screeds of such blossomings
    Should fill each lenten interval
    Lobster-clawed love should diminish
    On the roads leading to all countries
    Famine veers away

    They said maritime provinces
    N or M
    It isn't easy to see in this light
    And night writes no replies

MARROW

O talisman and all the rest
Where is the teeming myriad gone
I seem to see a mushroom growing upon the globe
Women are often spectral
They often walk down the street like banjos
Their eyes are often no more than mere scraps of paper

Incandescent mutability
Decrees that emotion goes early to bed
Metallic starshine of the mood
Indicates losing breath
Losing head and heart
In the shopwindows of the wind
Like watercress

Until I wear the close chaplet
There will be no more time for tears.

BAPTISM

Have had enough barbarity
But enough too of illusion
Dreams of peace

Walking in the water
Or upon it
With wet fingers on the brow
And sombre eyes turned upwards
No longer expectant but prepared
Have had enough of was . . .

Statement:
If you are with us you are red.

THE END IS NEAR THE BEGINNING

Yes you have said enough for the time being
There will be plenty of lace later on
Plenty of electric wool
And you will forget the eglantine
Growing around the edge of the green lake
And if you forget the colour of my hands
You will remember the wheels of the chair
In which the wax figure resembling you sat

Several men are standing on the pier
Unloading the sea
The device on the trolley says MOTHER'S MEAT
Which means *Until the end.*

LOST WISDOM

In the first morning
A cry above the unborn roofs
Of solitude and pain
A faint odour of vegetable matter
Fringing the violet lids of night
And hanging from the water's eyes
The simulacrum of the damned

Disturbance in the weather makes me see
The little angels without wings
The brittle needles in the sand
The ropy veins of polypi
And all the seamless seams

And now and then
From every abandoned mouth
An unstanched stream must flow
And then as now
The graves were opened once
And gold was melted by the snow
Like lilies sown in sifted stone
And gathered once for all.

DIRECT RESPONSE

The four elements are sitting at the table
There is a shipwreck on the sands
A warm hand in the mist
Flowers turn colour in the mist
Without moving

Sensitive needle at the extremity of breathing
What can you etch upon the eyes' quick web?
Up to your middle in the dewy grass
Whose profile can you sketch upon their filmy screen?

I have long forgotten why I am young
A bird's blue shadow trembles on my breasts
A bird's song blossoms from the water
Till my neck bends back in a curve like stone
And I am neither white nor warm nor cold

FUTURE REFERENCE

The roof-garden was full of strangled flowers
Full of stones like feet and feet like fronds
There was a still pool in the garden's eye
But now there is no more time to see
How in the unanimously carried vote of censure
There could be even a vestige of saltpetre
How could there be a voice talking in the annexe
How could there be a machine to reproduce trees
And if all these questions remain unanswered
It is not the fault of the cheese-mites
Those dainty creatures with fleur-de-lys on their breasts
No it is not my fault if the ovens get cold
Nor yours if the blades of the swords get warm
My little dog has folds of skin round his arse
That worry him all day long
My little Jesus my little Jesus what pretty curls you've got
What pretty pansies you wear in your seething hair
And if the oppressive odour of gelatine gets too much for you
Under your eyes will appear a whole hornets' nest
To tickle your weeping glands.

But stay but stay you have not yet learnt to fly
You cannot climb stairs without chains round your knees
Nor will the sky descend to kiss you
Till every aquarium under the sun is broken
And duets are danced no more in the holes in the sand
For colossal fruits are about to fall from the trees
And every child in the world will be able to bite their pink flesh
A colossal thigh covered with veins
Is the monument to be raised on the seaswept shore
To all who have lost their lives in pursuit of a dream.

AUTOMATIC ALBUM LEAVES

I

The room is not very large, plaster has begun to flake from the ceiling, the windows are draped with whorled lace, under the windows there are little orange trees growing in patent-leather shoes instead of pots. On the plain walls there are hundreds of crosses hanging, made of rotting, worm-eaten wood, and to each one is nailed a small flat figure cut from rose-coloured tin. On the table there are bundles of hair, paper-knives, photographs of angels kissing, bottles of chlorine, miniature facsimiles of the Discobolus in cork, and specimens of the handwriting of children. A tattered shadow floats upwards towards the ventilator and bursts there like a silent bomb; portfolios open on all the shelves and coloured plates showing embryonic development flutter down in slow-motion to the cement floor, faintly phosphorescent and smelling of sweat.

II

And did they never show you the heavenly respiration-box with its nine coagulated wounds and its ink-stained mouth into which they used to pour gall-stones? Did they never ask you to repeat the numbers and the corresponding colours of the seasons? I refer of course to the mental seasons, the spectacular Roman Catholic seasons, each one of which has its own cubicle and its own special electric comb. They would never have shown you the seasons' combs, they have to be kept apart during the day, as they are apt to turn black and to sprout poisonous feathers. It would never do to show them you were interested or that you had false teeth. Teeth are highly symbolic, and when dissolved in mercury they often re-solidify as algae with eyes as flowers, emitting a deadly violet aura.

31

Several years ago I fell violently in love with a pear tree, and sat for a long time in one of its branches. At the end of a week an enormous brown butterfly came flying towards me out of an oval mirror, circled four times round my head beating a little gong that was hidden under its wings, and ended tramping heavily for more than an hour upon my eyelids, while I repeated the words: 'dust to dust, ashes to ashes,' over and over again. Then I opened my mouth and spat out a large quantity of blood on to the ground. The butterfly disappeared, but several alarm-clocks with miniature pairs of claws and daffodil-heads instead of bells swam to the surface of the blood that I had spat out and began to climb up the trunk of the tree in which I was sitting. I was so terrified that I expectorated loudly several times, but my fears were groundless for the alarm-clocks did not come near the branch on which I had made my perch, they climbed without pausing to the very topmost bough, where they astonished birds by committing hara-kiri, en masse.

IV

I can see a how-d'you-do in the inkbottle. It will taste slightly of lemon and it will enable you to balance cameras on your little finger. Be careful not to release the third spring, that is the one that poisons dowagers and makes its bed in latrines. And be careful not to look at the weathercock, as its eyes have been replaced by snowballs containing the stamens of certain plants that are extremely dangerous to the human eye between half-past two and three in the morning. When you reach the garage with the dead bird hanging up outside, turn to the left and go down the alleyway. Half-way down the alley you will meet a naked woman who will take you by the hand and drag you out onto the esplanade. There you will receive the secret message about which I have already told you.

V

How historic and full of resonance are these crumbs! With their delightful bell-shaped dresses and their little heads like daggers they are capable of performing all kinds of aquatic tricks, such as high-life bicycling and shoe-blacking, and they are able to peer down at leisure into the great globe that stands in the garden. That they are energetic and truthful I can only guess, but that their music is sweeter than parsley basins or mesmeric influences I am certain. Observe the way they glide out of the churches, lifting their skirts with lobster-like delicacy, watch how carefully they transform themselves into Japanese plants and begin to scrub the floor of the travel bureau! Every moment would be disastrous if they took no precautions against breathing, but happily for us they have a deeply-ingrained sense of gratitude to the human race. They will never risk thinking in the ordinary, mediocre and *sensible* way that people like ourselves do!

VI

Suffering has abundant hair and lives alone on a cliff. The walls of its house are round and they move up and down like birds' eggs in a mirage. Bunches of smoke like umbrellas live nearby, with trembling fingers like glass keys on each of which some date famous in history has been written. They shake their rattling wrists and Suffering flies up the chimney and opens its green chemical head like a work-box. All the trout that swim continuously through its hair begin to murmur ominously and to turn on and off, their electric danger signals. Their teats begin to spurt gin and the gnats that Suffering keeps in the little wire cage under its left shoulder-blade lap up the gin thirstily. Promiscuous miracles flash through the air and set light to the balancing wardrobes. Armies of pigtails march past, winking and playing bridge. Priests pull pieces of string through holes bored in their jaws. Nursemaids paint their charges' behinds with glue. And at the supreme moment of universal anguish, when all the cathedrals are wrapped in blue cotton-wool, and all the underground railways are flooded and when trees with diseased branches have burst upwards through the pavements of the great cities, and choked every building with dense, verminous foliage, at the supreme moment Suffering passes right through the centre of the bunches of smoke, causing a revolting stench of burning rubber, and appears on the other side dressed in white, wearing a bishop's mitre stained with ink, wine, blood and sperm.

VII

The little pilgrims that make their nests in stones are afraid of the wings of swallows. They are afraid of street-corners.

A bow and arrow appears on the edge of a cloud, while an iron figure eight slides up the wall. Your ear opens like a window-shutter and inside I can see rows and rows of little bells arranged in a tea-leaf pattern. I tell you again and again: *'There is a crystallized hair in the last workman's bouquet and fish are cleaning their nails with the rose-coloured pencils of despair.'* But you cannot hear me; all the little bells in your head are playing The Bluebells of Scotland.

33

The little pilgrims are standing at the corner of the street, shadowed by fluttering wings. They are shouting at the tops of their voices.

VIII

My eyelashes are as big as churches. It makes me very happy to be able to read the word ROME on the top of your stocking. Please do not stare at my lips, I know they are swollen, it is because of the wooden doll that I use for brushing my teeth. If you like I will show you a photograph of a clergyman; he is said to have made advances to the statuette of the Madonna that he keeps in his study.—You do not like madonnas? Neither do I; they are made of pig's fat and their haloes have an unpleasant smell. Let us amuse ourselves by tearing to pieces this shirt-front with consecrated buttons.

When night falls, a great many sins will be committed in the grove by the holy well.

IX

It is not fashionable to wear zinc next the skin. In any case, most people prefer goats' feathers. And Sappho sings enchanting sea shanties in the café next door in a next-to-nothing blouse. How can I keep her waiting?

It is beginning to rain now and I have at least seven trains to catch. If there should be an explosion, would you be prepared to rescue the time-table? It is in the wash-basin. Be careful of the telescope just inside the door.

Does anyone know how to play conkers, a recipe for bats in the belfry, the way to lick stamps, the way to London; how to open collection boxes?

THE LIGHT OF THE LION'S MANE

If I had a candle I would bite it in half, avid with spite and angry greed. Why is it that candles give no light unless they are wrapped in oak-leaves that have been pressed between the pages of books illuminated by ancient monks? I hate to see them melting away like that, losing first their heads and then their tails, and balancing what remains of them as a fishmonger balances his wares on top of a pile of wicker baskets. It would be better to send for a pound of butter on a chafing-dish. Butter is far better, for it does not bite the tongue.

The violet light thrown by the lamp strapped on to the miner's forehead falls gently on to the surface of a subterranean river. Eagles have made their nests along the banks, and the fossilized claws of the neolithic tiger are to be found in the rusty sand of the river's bed. I often sleep there, and when I wake up it always seems as though a procession of foreign tourists had passed by that way during the night, gesticulating with their arms and making lengthy speeches at every turn of the twilit tunnel. I should never have been able to explain to them why the tattered curtains are alive with toads, or why the spectacular staircase is in ruins. It is a long time since I spat into so deep a hole; I have never seen such pilgrims, with their bells, their books, their baskets . . .

But let us pause to consider the cause of the disturbance that is taking place at the far end of this corridor. The waves have thrown up the remains of a small vessel on to the sanded floor and among the shattered casks and the crumbs of the ship's biscuits one can see a dissevered head that is trying to speak to the assembled multitude. The muscular effort made by its jaw is equal to ten times the strength of a derrick trying to break away from the crane to which it is tethered. It is covered with sunspots and will undoubtedly burst into flame at the end of a quarter of an hour or so. Lay your hand on the massive forehead and you will feel the gradual movement of the birds that are imprisoned underneath. Each bird carries a leather glove in its beak, and the fingers of each glove are packed with gunpowder. The final explosion has been timed to coincide with the demolition of the plaster-of-paris monument that has been set up in the middle of the park to commemorate the victims of a savage watchdog who wrought great ravages in these parts towards the end of the nineteenth century.

Heads such as these do not speak as clearly as the heads of missionaries. Let us set up a temple for the Alpine mission, and let us weave a great carpet at the foot of every mountain in the Alps, to express our penitence and our desire to make amends for the broken glaciers and the training clouds of glory. Our whole childhood was spent in the shadow of these great heights, so is it not only right that we should decorate them now with dazzling garments stained with our own blood? I have often expressed a desire to lie down on the floor of a cave, and it seems that my wish is at last about to be granted. Have I put on my head-dress straight?

Looming out of the gloomy shadows of the further chamber there comes a great catafalque drawn by a pair of milk-white does and decked with plumes of lilies and clustering branches of tiger-lilies that look like sword-lilies. It has almost the appearance of a November bonfire set alight in the public square because of a plot that failed. The lights are turned on one by one, the leaves of the candelabrum-trees are shining like buttered gold, the foam of the Gulf Stream glitters like corn in the sun, and the whole effect is one of heat, drums and fireworks. The monster Egg that forms the centrepiece of these celebrations now bursts open, and a living Archangel leaps out. Nine months ago she was but an atom whirling through the wastes of outer space, and now her robe is bright with sweat and all eyes are turned towards her. She blows one blast on her vast brass trumpet shaped like an oar, and the whole brilliant pageant falls to dust.

But who has tied a bandage round my eyes so that I can no longer see what is happening? A bandage saturated with the scent of crushed laurel-leaves, which is used by butterfly-hunters. I have the sensation of being driven away in a rickshaw, I could swear that I heard foot-steps behind me, the wheels of the conveyance bump loudly down the stairs. Clusters of sharp little shells are growing beneath my eyelids, ants' eggs to throw to the fishes, chrysalises lying quietly in the dust beneath the feet of the marching tyrants, who will all fall down with fatigue in the end, and bury their arrogant faces in the mire.

THE GREAT DAY

When I woke up it was indeed very beautiful. The banisters were shining intensely and the stairs were coming up towards me. I was well aware that my eyes were no longer clinkers. I sat on the edge of the bed with my feet in the sand and watched the ambulances going past the window. What carnage, what thunderbolts and, indeed, what pascal lambs!

But I'm afraid you will hardly believe me when I tell you that at the hour when the night-bird should have flown, at the hour when all the matrons no longer able to have children should have entered the room, precisely at the hour of the one-o'clock séances and balloon-course meetings, it was one o'clock. I went out as the cock was crowing and held my head above the basin which I thought was full of water but it was full of cream and ashes. This, of course, brought on one of my fainting fits, but I soon recovered, and **there**, to my infinite surprise, sitting upon the left-hand flap of the little linoleum wigwam which looks like a forge-bellows, was she upon whom my heart had been set ever since that marvellous sunset long long years ago when my heart was still a captive beating its pitiful wings in the great silence of all the empty rooms and the dining-rooms and the cellars and all the wine-cellars. Without a moment's hesitation I went straight up to her and caught hold of her icy hand, I can tell you, and her mouth was like a beautiful garden full of flowers and full of bronze flowers and beautiful flowers like medals. My adoration knew no bounds and the sound of my kisses on the air was like the flapping of sheets, I know what I am saying, it was like the bottling of new wine. But what was my amazement and despair when she told me she could never be mine for she was married to a leper, imagine it, what could I do to prevent my heart from bursting into a million little pieces like diamonds and emeralds and rubies, yes real ones, not imitation glass ones, never, I have never stooped to that. She tore her hands and feet away and a great pain shot through me like a shaking spear, for it was she who had taught me all those wonderful words, it was she whose blood I had wanted to feel pulsing beneath mine own, and now she refused to open her veins for me! My passion was so frightful that I might have spat right in her face, but fortunately I was able to restrain myself and she passed away like the great wave after the earthquake of Messalina.

When everything was once more as clear and as peaceful as the falling rain and the terrible burden of my sighing had lifted itself from my poor ravaged breast, I was able to see all the dear little children playing at blind hands' muff on the mantelpiece. I took out my great burnished watch that sings like a bird and whose very hands are like feathers and whose face is divided into four sections that are the four seasons all coloured like the rainbow. I even went so far as to open it for them and show them all the needlebones and chalcedonies going round and round in its chemical inside. I take a great delight in mechanisms of all kinds, especially those that repeat themselves like the famous reproduction of the great hunting-horn that hangs on the wall in my family ward.

And then it was at last time for the operation. Were I to describe to you all the details of what took place on that memorable occasion it would take me ten times as many books as there are stars in the universe and in any case my pen would have turned to dust long before I got to the last astonishing page where I should sign my name in letters of flame and of gold and in letters of flaming gold.

First of all it was like **drinking oxygen**. I had the gentle maternal pigeon on the one side of me and the symbol of the crossed keys on the other, so I felt perfectly safe. It was like looking at that picture of a girl

37

climbing a rope which hangs on the wall in the warden's room, it was like woollen buttons and angel's skin. It kept changing all the time, of course, so that one minute you saw the pattern of the minutes coming and going and the next you saw the sort of sawdust that they throw down on the floor if you look at it hard enough. I stretched out my hands and they went sliding far away out over the multitudinous seas whose voices came to me like the sound of chariots and firearms roaring and terrible chariots grinding the limbs of the helpless Christians to powder. Then the bed started to go up and down but it wasn't a bed it was a sort of automatic pianola and it began to gallop away with me on its back right into the middle of the forest where the chimneys were all smoking away like fury because the silly things thought it was the middle of the night. But I knew better, of course, so I sat up there and then and told them that I wasn't going to stand any more of it and I smote the ridiculous creature with the wooden leg a terrific blow across the backside, and they were all absolutely terrified of my voice like hundreds of railways thundering and my face like a red indian's.

But what am I saying? They thought they could scratch me with their tigers' claws and their eagles' talons, the wretches, they thought they could scratch my eyes out, but they weren't going to get away with it so easily. I lifted my imperious iron hand, I whose hands and feet are the very seal of all that is powerful and triumphant in this miserable world where the flowers only grow to please me, I lifted my iron hand and it became a sword and sceptre against all the wicked and unruly tongues that were clacking in the caverns in the valley of the shadow of death. My breathing became like the wind of the great tempest and I felt my body growing to stupendous size and the blinding light was like organs playing. What noble pity surged into my melting soul and how I knew everything that had been forgotten down the centuries by the mages and the saviours and the nobility of all European countries! For that was easily the greatest moment of all, when all the candles were being burnt for me and all the banquets were being given in my honour and all the assembled nations were singing songs in which my name was mentioned at least once, I think I might even say without boasting that it was mentioned ten times, in every verse.

After that, as you will well understand, it was not so difficult for me to come back into the daylight. The room was just the same as before except that the window seemed to have lost something of its original transparency and the table had been replaced by a milk-float. Nobody seemed to notice any particular change in my appearance, but if they had looked closely enough they could not have helped seeing the little snow-white footprints on my eyelids and the little black stars on my lips. In any case I took no notice of them, for I despise all men who have not the words LOVE AND DEATH inscribed on their banners, and when I went out in the evening I met my mother walking in the garden. She was wearing one of my most cherished hats and I told her of all my recent experiences, ending up by explaining how I had been awarded the Nobel Peace Prize for my exploits among the redskins. She smiled gently and, lifting her veil, began to talk about the time when she went to tea with George Sand. Then we went to choose the flowers for the wreath. And the phosphorescent night began to fall.

Night, yes indeed it was the night that fell, for I distinctly saw its columns dissolving one into the other and its arches falling and its great **aqueducts** falling down like the very symptoms of a weak heart after taking belladonna. I knew it was soon going to be very beautiful again and I was just as sure that, after what had been revealed to me and to me alone, I should never fall down. Two very massive and indestructible shoulders support this noble and imposing head of mine, this head which is so full of gorgeous pictures of the wonderful palaces, castles, fortresses and great endless glittering palaces that are my inheritance and where I shall at last rest these weary bones of mine, far from the stupid creatures I despise, far from the snaggle-toothed turnip-heads and the heartless women whom I still adore although they have made my life such a misery, far, I say, from the turnip-tops and the butterflies' hearts and the rascally curly-locked gas-meters.

And now it is time for me to end, or rather, since I never really **end**, shall I say **come to** an end before saying good night to you and a downright sentimental journey.

LOZANNE

It was seven, it was nine o'clock, the doors were closing, the windows were screaming. You bent over the shadow that lay on the floor and saw its eyes dissolving. The band about your forehead began to turn. The band of fever.

The armchair turned into a palace, the carpet became a bank of withered flowers, *and then it was time to go.* Every semblance of that which had gone before became the means by which you ascended the great staircase. And took your place among the stars.

For it is significant, is it not, that the *blemish* about which you were so insistent was nothing less than that interminable voice which haunted you in your dreams, saying 'I love you' over and over again. And the panelling of the room where they asked you questions was made of exactly the same wood as the mallet which you had to hate.

The dusty and ashen residue of a passion that now raged elsewhere, but still raged, rose slowly upwards to the surface of the lake as your blood sank slowly through it. And the other returned to ice. Oh, I can see through your eyes now and I can see what flame it was that melted everything before it! (Though the obstinate sod refused to become softened by the rain of thaw.) But you were spared passing through that black box where a masked man kisses his victim before her death. I ask the glass again: Who gave the victims right to refuse life to those who refuse to be victimized?

Those who damned shall be damned.

PHENOMENA

It was during a heatwave. Someone whose dress seemed to have forgotten who was wearing it appeared to me at the end of a pause in the conversation. She was so adorable that I had to forbid her to pass across my footstool again. Without warning, changing from blue to purple, the night-sky suffered countless meteoric bombardments from the other side of the curtain, and the portcullis fell like an eyelid.

The milk had turned sour in its effort to avoid the centrifugal attraction of a blemish on its own skin. Everything was mounting to the surface. My last hope was to diminish the barometric pressure at least enough to enable me to get out from beneath it alive.

In the end, I remembered that she would not have to make the decision herself, as her own fate was sufficient justification for the hostility of the elements. I turned the page. Nothing could have been more baffling than the way in which the words rose from the places where they had been printed, hovered in the air at a distance of about six inches from my face and finally, without having much more than disturbed my impression of their habitual immobility, dissolved into the growing darkness. As I have said, it was during a heatwave, and the lightning had well nigh worn itself out in trying to attain the limit of its incandescence. I suddenly forgot what I was supposed to be doing, and the soil beneath my feet loosened itself from the hold of the force of gravity and began to slide gradually downwards, with the sound of a distant explosion.

REFLECTED VEHEMENCE

Umbilically detached, of sorrowful mien and at the same time decked out in cobwebs—these vanquished ones, whose breathings propagate violence and fear. Their padded fingers point uselessly to the stars of their own eloquence. It is just the same as ever in the outgrown pavilions of vegetable matter. As though St Valentine had smudged the last letters of a secret pact with the powdered antennae of a forgotten fly. As though flying itself were only circular.

But here where the graphite byways meet, there is bound to be always fresh water. See how the ruched waterfalls reply with shaking heads to the invitations of the warrior-like foliage. They seem to vanish in thin air, gasping for a more fluid means of expression. The tinkling belfries glide away of their own volition. Eggs break during the fencing lesson. Masonry, tightly clamped to the nape of the ritual, buries itself in an indulgently frothing explosion of the head, whereby the closed gates are breathed upon anew by the breezes of loyalty and honour. Thus clouds are born.

In my hand lies the same whispering, nail-headed dude, ever imploring the benefice of a hippograph.

NO SOLUTION

Above and below
The roll of days spread out like a cloth
Days engraved on everyone's forehead
Yesterday folding Tomorrow opening
Today like a horse without a rider
Today a drop of water falling into a lake
Today a white light above and below

A fan of days held in a virgin hand
A burning taper burning paper
And you can turn back no longer
No longer stand still
The words of poems curling among the ashes
Hieroglyphics of larger despairs than ours.

UNSPOKEN

Words spoken leave no time for regret
Yet regret
The unviolated silence and
White sanctuses of sleep
Under the heaped veils
The inexorably prolonged vigils
Speech flowing away like water
With its undertow of violence and darkness
Carrying with it forever
All those formless vessels
Abandoned palaces
Tottering under the strain of being
Full-blossoming hysterias
Lavishly scattering their stained veined petals

In sleep there are places places
Places overlap
Yellow sleep in the afternoon sunlight
Coming invisibly in through the pinewood door
White sleep wrapped warm in the midwinter
Inhaling the tepid snow
And sleeping in April at night is sleeping in
Shadow as shallow as water and articulate with pain

Recurrent words
Slipping between the cracks
With the face of memory and the sound of its voice
More intimate than sweat at the roots of the hair
Frozen stiff in a moment and then melted
Swifter than air between the lips
Swifter to vanish than enormous buildings
Seen for a moment from the corners of the eyes

Travelling through man's enormous continent
No two roads the same
Nor ever the same names to places
Migrating towns and fluid boundaries
There are no settlers here there are
No solid stones

Travelling through man's unspoken continent
Among the unspeaking mountains
The dumb lakes and the deafened valleys
Illumined by paroxysms of vision
Clear waves of soundless sight
Lapping out of the heart of darkness
Flowing endless over buried speech
Drowning the words and words

And here I am caught up among the glistenings of
Bodies proud with the opulence of flesh
The silent limbs of beings lying across the light
Silken at the hips and pinched between two fingers
Their thirsty faces turned upwards towards breaking
Their long legs shifting slanting turning
In a parade of unknown virtues
Beginning again and beginning
Again

Till unspoken is unseen
Until unknown
Descending from knowledge to knowledge
A dim world uttering a voiceless cry
Spinning helpless between sleep and waking
A blossom scattered by a motionless wind
A wheel of fortune turning in the fog
Predicting the lucid moment
Casting the bodiless body from its hub
Back into the cycle of return and change
Breathing the mottled petals
Out across the circling seas
And foaming oceans of disintegration
Where navigate our daylight vessels
Following certain routes to uncertain lands

THE LAST HEAD

In the warm sand-coloured room at the end of the watery road
I saw the last head with its fingers plaited in curls
And its sides ridged and smooth, worn by runnels of light.
The obvious table supported a map of the moon.

The faces in trees must be stopped, and the towers
And peninsular madness and gems
The canals are all stopped with a white-flowering weed
The beetle conspires to bring doom to the bridge
The night air is salt on the tongue. The white shields
In the stable fall clattering down from the walls.

But the last head is safe in its vegetable dome:
The last head is wrapped in its oiled silk sheath,
While the pale tepid flame of its ichorous brain
Consumes all its body's dry shells.

PURIFIED DISGUST

An impure sky
A heartless and impure breathing
The fevered breath of logic
And a great bird broke loose
Flapping into the silence with strident cries
A great bird with cruel claws

Beyond that savage pretence of knowledge
Beyond that posture of oblivious dream
Into the divided terrain of anguish
Where one walks with bound hands
Where one walks with knotted hair
With eyes searching the zenith
Where one walks like Sebastian

Heavy flesh invokes the voice of penitence
Seated at the stone tables
Seated at a banquet of the carnal lusts
Behind our putrid masks we snicker
Our men's heads behind our masks
Twisted from innocence to insolence

And there the pointing finger says and there
The pointing finger demonstrates
The accuser struggles with his accusation
The accused writhes and blusters
The finger points to the chosen victim
The victim embraces his victimization
The accused belches defiance

How could we touch that carrion?
A sudden spasm saves us
A pure disgust illumines us
The music of the spheres is silent
Our hands lie still upon the counterpane
And the herds come home.

CHARITY WEEK
To Max Ernst

Have presented the lion with medals of mud
One for each day of the week
One for each beast in this sombre menagerie
Shipwrecked among the clouds
Shattered by the violently closed eyelids

Garments of the seminary
Worn by the nocturnal expedition
By all the chimeras
Climbing in at the window

With lice in their hair
Noughts in their crosses
Ice in their eyes

Hysteria upon the staircase
Hair torn out by the roots
Lace handkerchiefs torn to shreds
And stained by tears of blood
Their fragments strewn upon the waters

These are the phenomena of zero
Invisible men on the pavement
Spittle in the yellow grass
The distant roar of disaster
And the great bursting womb of desire.

YVES TANGUY

The worlds are breaking in my head
Blown by the brainless wind
That comes from afar
Swollen with dusk and dust
And hysterical rain

The fading cries of the light
Awaken the endless desert
Engrossed in its tropical slumber
Enclosed by the dead grey oceans
Enclasped by the arms of the night

The worlds are breaking in my head
Their fragments are crumbs of despair
The food of the solitary damned
Who await the gross tumult of turbulent
Days bringing change without end.

The worlds are breaking in my head
The fuming future sleeps no more
For their seeds are beginning to grow
To creep and to cry midst the
Rocks of the deserts to come

Planetary seed
Sown by the grotesque wind
Whose head is so swollen with rumours
Whose hands are so urgent with tumours
Whose feet are so deep in the sand.

SALVADOR DALI

The face of the precipice is black with lovers;
The sun above them is a bag of nails; the spring's
First rivers hide among their hair.
Goliath plunges his hand into the poisoned well
And bows his head and feels my feet walk through his brain.
The children chasing butterflies turn round and see him there
With his hand in the well and my body growing from his head,
And are afraid. They drop their nets and walk into the wall like
  smoke.

46

The smooth plain with its mirrors listens to the cliff
Like a basilisk eating flowers.
And the children, lost in the shadows of the catacombs,
Call to the mirrors for help:
'Strong-bow of salt, cutlass of memory,
Write on my map the name of every river.'

A flock of banners fight their way through the telescoped forest
And fly away like birds towards the sound of roasting meat.
Sand falls into the boiling rivers through the telescopes' mouths
And forms clear drops of acid with petals of whirling flame.
Heraldic animals wade through the asphyxia of planets,
Butterflies burst from their skins and grow long tongues like plants,
The plants play games with a suit of mail like a cloud.

Mirrors write Goliath's name upon my forehead,
While the children are killed in the smoke of the catacombs
And lovers float down from the cliffs like rain.

THE VERY IMAGE
To René Magritte

An image of my grandmother
her head appearing upside-down upon a cloud
the cloud transfixed on the steeple
of a deserted railway station
far away

An image of an aqueduct
with a dead crow hanging from the first arch
a modern-style chair from the second
a fir tree lodged in the third
and the whole scene sprinkled with snow

An image of the piano tuner
with a basket of prawns on his shoulder
and a firescreen under his arm
his moustache made of clay-clotted twigs
and his cheeks daubed with wine

An image of an aeroplane
the propeller is rashers of bacon
the wings are of reinforced lard
the tail is made of paperclips
the pilot is a wasp

An image of the painter
with his left hand in a bucket
and his right hand stroking a cat
as he lies in bed
with a stone beneath his head

And all these images
and many others
are arranged like waxworks
in model birdcages
about six inches high.

'THE TRUTH IS BLIND'

The light fell from the window and the day was done
Another day of thinking and distractions
Love wrapped in its wings passed by and coal-black Hate
Paused on the edge of the cliff and dropped a stone
From which the night grew like a savage plant
With daggers for its leaves and scarlet hearts
For flowers—then the bed
Rose clocklike from the ground and spread its sheets
Across the shifting sands

Autumnal breath of mornings far from here
A star veiled in grey mist
A living man:

The snapping of a dry twig was his only announcement. The two
men, who had tied their boat to a branch that grew out over the water's
edge, and were now moving up through the rank tropical vegetation,
turned sharply.

He raised his eyes and saw the river's source
Between their legs—he saw the flaming sun
He saw the buildings in between the leaves
Behind their heads that were as large as globes
He heard their voices indistinct as rain
As faint as feathers falling
  And he fell

The boat sailed on
The masts were made of straw
The sails were made of finest silken thread
And out of holes on either side the prow
Gushed endless streams of water and of flame
In which the passengers saw curious things:

The conjuror, we are told, 'took out of his bag a silken thread, and
so projected it upwards that it stuck fast in a certain cloud of air. Out of
the same receptacle he pulled a hare, that ran away up along the
thread; a little beagle, which when it was slipped at the hare pursued it
in full cry; last of all a small dogboy, whom he commanded to follow
both hare and hound up the thread. From another bag that he had he
extracted a winsome young woman, at all points well adorned, and
instructed her to follow after hound and dogboy.'

She laughed to see them gazing after her
She clapped her hands and vanished in thin air
To reappear upon the other bank
Among the restless traffic of the quays
Her silhouette against the dusty sky
Her shadow falling on the hungry stones
Where sat the pilot dressed in mud-stained rags

He knocked the fragile statue down
And ate her sugar head
And then the witnesses all gathered round
And pointed at the chasm at his feet:

Clouds of blue smoke, sometimes mixed with black, were being
emitted from the exhaust pipe. The smoke was of sufficient density to
be an annoyance to the driver following the vehicle or to pedestrians.
  The whispering of unseen flames
  A sharp taste in the mouth.

THE CAGE

In the waking night
The forests have stopped growing
The shells are listening
The shadows in the pools turn grey
The pearls dissolve in the shadow
And I return to you

Your face is marked upon the clockface
My hands are beneath your hair
And if the time you mark sets free the birds
And if they fly away towards the forest
The hour will no longer be ours

Ours is the ornate birdcage
The brimming cup of water
The preface to the book
And all the clocks are ticking
All the dark rooms are moving
All the air's nerves are bare.

Once flown
The feathered hour will not return
And I shall have gone away.

EDUCATIVE PROCESS

I

What though the weather changes?
What though you do not sleep?
Now that at last we've arrived
(Forget the wasteproducts of love)
Whiteness envelopes houses
To prepare to begin to prepare
· And snow on the roofs,
Your horror of snow!

2

The month's pocket holds many days
The paraphernalia of seeing and hearing.

3

The feathers fledged from your flesh meet mine
And ardent haloes meet like plates above our heads

You are not gentle.

4

Crescendo of flames, the steps
Of stone that lead into the swamp
Where wanderings begin and the first birds
The last birds, the sun's bicycle racing,
Our eyes lose one another, autumn splutters
On the sidewalks houses eat the afternoon
Soft outline of the leaves upon the wall
Foliage blown by the wind
Streams into the memory of hair.

5

Wire twisted back bites into the cheek
The gardens of neurosis.

6

Swift algebra of love pretends
That barriers must fall
To gourmandize the warriors of sleep
To sacrifice the carrion
To call home lightnings wandering in the fields
To live life twice.

7

A drop of dew sings psalms upon the hill
Anatomies of wonder opened at the first page
The last page showing the number 3 like a silken knot.

Rockets open the sky like keys
And your breath is warning
Warning the footsteps of Truth
Not to wander too far away
For clutching hands and agonized eyes
Move with their shadows upon the imaginary screen.

Hooped foliage, tired antimony,
Blossoms of crumbling columns beneath our feet
Journeys stretch far away and there is the sea
The sea is as salt as health with its marble veins.

9

The glass on the table is empty and so are your eyes.
Footsteps. The shadow just outside the door.
And do you suppose that forgetting
Is as easy as air?

The flowers' voice is evil, the caves
Are asleep. In the grass
Children playing take fear at the clouds carved like skulls.

10

I had forgotten to watch the wind
The wind playing with boats the wind
Shuffling the sands like cards—
But we cannot change now that daylight is here

Negotiations with the infinite
Upon the empty beach.

ANTENNAE

1

A river of perfumed silk
A final glimpse of content
The girls are alone on the highroad.

2

In the evening there is a cry of despair
Silence begins spawning its myriad
Shifting away from the restless neon auras
Disturbed by the menacing gestures of starvation
The unchanging programme of its manoeuvres
Its rasping grasping claws.

3

The sun bursts through its skin
The last smooth man emerges from the tunnel
And flags burst into song along the streets
The morning's garlands pull themselves to pieces
And fly away in flocks

The sea is a bubble in a cup of salt
The earth is a grain of sand in a nutshell
The earth is blue.

4

Truth, fickle monster, gazed in at the open window
Longing to eat of the fruit of the poisoned tree
Longing to eat from the plates on our lozenge-shaped table
Fearing the truth

And the peaceful star of the vigil fell from the sky
And spilt its amazing fluids across the mosaic floor.

5

The timeless sleepers tangled in the bed
In the midst of the sonorous island, alone

The tongue between the teeth
The river between the sands

Love in my hand like lace
Your hand enlaced with mine.

6

A delicate breath a wisp of smoke
Floating between our eyes
The rainbow-coloured barque of pleasure
Brushing the fluid foliage aside
Derision's flimsy feathers

Between our eyes
The shadow of a smile.

7

The full breasts of eternity awaiting tender hands.

8

Not wholly unprepared
Nor entirely unafraid
Vigilant
Watching the colours

Discovered by morning:
Dispensation of doubtful benefits.

9

At least alone at last
When gone the body's warmth
The incisiveness of glances
The unwinding crimson thread
The given flower

Forgotten mouths forget.

10

For now we are suspended above life
There are a great many questions to be answered
A great many debts to be paid

So evanescent that which binds us
That more is meant, regret is absent . . .
Our burning possession of each other
Held in both hands because it is all we have.

THE DIABOLICAL PRINCIPLE

The red dew of autumn clings to winter's curtains
And when the curtain rises the landscape is as empty as a board
Empty except for a broken bottle and a torso broken like a bottle
And when the curtain falls the palace of cards will fall
The card-castle on the table will topple without a sound

An eye winks from the shadow of the gallows
A tumbled bed slides upwards from the shadow
A suicide with mittened hands stumbles out of the lake
And writes a poem on the tablets of a dead man's heart
The last man but one climbs the scaffold and fades into the mist

The marine sceptre is splintered like an anvil
Its spine crackles with electric nerves
While eagle pinions thunder through the darkness
While swords and breastplates clatter in the darkness
And the storm falls across the bed like a thrice-doomed tree.

                    *

    A basket of poisoned arrows
    Severing seawrack, ships' tracks
    Leadentipped darts of disaster
    A unicorn champs at the waves
    The waves are green branches singing
    The cry of a foal at daybreak
    A broken mouth at sunset
    A broken lamp among the clouds' draperies

    A sound drops into the water and the water boils
    The sound of disastrous waves
    Waves flood the room when the door opens
    A white horse stamps upon the liquid floor
    The sunlight is tiring to our opened eyes
    And the sand is dead
    Feet in the sand make patterns
    Patterns flow like rivers to the distant sky
    Rippling shells like careful signatures
    A tangled skein of blood

    In fumigated emptiness revolves the mind
    The light laughs like an unposted letter
    Railways rush into the hills.

                    *

A worm slithers from the earth and the shell is broken
A giant mazed misery tears the veil to shreds
Stop it tormentor stop the angry planet before it breaks the sky

Having shattered the untapped barrel
Having given up hope for water
Having shaken the chosen words in a hat
History opened its head like a wallet
And folded itself inside.

THE RITES OF HYSTERIA

In the midst of the flickering sonorous islands
The islands with liquid gullets full of mistletoe-suffering
Where untold truths are hidden in fibrous baskets
And the cold mist of decayed psychologies stifles the sun
An arrow hastening through the zone of basaltic honey
An arrow choked by suppressed fidgetings and smokey spasms
An arrow with lips of cheese was caught by a floating hair

The perfumed lenses whose tongues were tied up with wire
The boxes of tears and the bicycles coated with stains
Swam out of their false-bottomed nests into clouds of dismay
Where the gleams and the moth-bitten monsters the puddles of soot
And a half-strangled gibbet all cut off an archangel's wings
The flatfooted heart of a memory opened its solitary eye
Till the freak in the showcase was smothered in mucus and sweat

A cluster of insane massacres turns green upon the highroad
Green as the nadir of a mystery in the closet of a dream
And a wild growth of lascivious pamphlets became a beehive
The afternoon scrambles like an asylum out of its hovel
The afternoon swallows a bucketful of chemical sorrows
And the owners of rubber pitchforks bake all their illusions
In an oven of dirty globes and weedgrown stupors

Now the beckoning nudity of diseases putrifies the saloon
The severed limbs of the galaxy wriggle like chambermaids
The sewing-machine on the pillar condenses the windmill's halo
Which poisoned the last infanta by placing a tooth in her ear
When the creeping groans of the cellar's anemone vanished
The nightmare spun on the roof a chain-armour of handcuffs
And the ashtray balanced a ribbon upon a syringe

An opaque whisper flies across the forest
Shaking its trailing sleeves like a steaming spook
Till the icicle stabs at the breast with the bleeding nipple
And bristling pot-hooks slit open the garden's fan
In the midst of the flickering sonorous hemlocks
A screen of hysteria blots out the folded hemlocks
And feathery eyelids conceal the volcano's mouth.

THE CUBICAL DOMES

Indeed indeed it is growing very sultry
The Indian feather pots are scrambling out of the room
The slow voice of the tobacconist is like a circle
Drawn on the floor in chalk and containing ants
And indeed there is a shoe upon the table
And indeed it is as regular as clockwork
Demonstrating the variability of the weather
Or denying the existence of manu altogether
For after all why should love resemble a cushion
Why should the stumbling-block float up towards the ceiling
And in our attic it is always said
That this is a sombre country the wettest place on earth
And then there is the problem of living to be considered
With its vast pink parachutes full of underdone mutton
Its tableaux of the archbishops dressed in their underwear
Have you ever paused to consider why grass is green
Yes greener at least it is said than the man in the moon
Which is why
The linen of flat countries basks in the tropical sun
And the light of the stars is attracted by transparent flowers
And at last is forgotten by both man and beast
By helmet and capstan and mermerised nun
For the bounds of my kingdom are truly unknown
And its factories work all night long
Producing the strongest canonical wastepaper-baskets
And ant-eaters' skiing-shoes
Which follow the glistening murders as far as the pond
And then light a magnificent bonfire of old rusty nails
And indeed they are paid by the state for their crimes
There is room for them all in the conjuror's musical-box
There is still enough room for even the hardest of faces
For faces are needed to stick on the emperor's walls
To roll down the stairs like a party of seafaring christians
Whose hearts are on fire in the snow.

THE SYMPTOMATIC WORLD

I

At the age of nine months I entered the world
As an automatic apprentice
My wages were divided
By the comparison between fire and water
My muscles were contracted
By the song and the wedding-ring
By the man in the front room smoking a cigar
And my eyes were especially opalescent
In that I gave them tears to drink each morning
Tears of warm milk in which flies were seen to float
Tears of cold amber in which miracles appeared
So that I seemed to see through them a world of metal
A world of intrinsic gestures and straight lines
I might even say a world in which there was no absence
And no unknown degrees
In which the pale green torture of the mountains
Appeared to consist of feathers sprouting from maps
And where the only women
Were negresses with breasts like collar-bones
And heads like violins played on by lightning
A world at last as empty as my mirror
Yet full of coach-horses and sails of ships
And vocal clocks all calling:
This way home.

II

Following an arrow
To the boundaries of sense-confusion
Like the crooked flight of a bird
The glass-lidded coffins are full of light
They displace the earth like the weight of stones
Eating and ravaging the earth like moths
Which follow the arrow
In a shower of freshly variegated sparkles
Confusing the issue of the arrow's flight
Till its feathers are all worn out
And the trees are all on fire
The pillow-case is bursting
The feathers are blown across the roofs
The room is falling from the window

And O where did that woman come from
Who chases the muleteer across the pampas
And covers her flaming face with the huge shadows of her hands?

III

The pinecone falls from the sailor's sleeve
The latchkey turns in the lock
And the light is broken
By the angry shadow of the knave of spades
Kneeling to dig in the sand with his coal-black hands
His hair is a kite to fly in the dangerous winds
That come from the central sea
He is searching for buried anvils
For the lost lamps of Syracuse
And behind him stands
The spectre whose lips are frozen
Unwinding the threads of her heart
From their luminous spool
She is stone and mortar
And tar and feather
Her errand is often obscure
But she comes to sit down in the glow of the rocks
She comes with a star in her mouth
And her words
Are rock-crystal molten by thunder
Meteors crushed by the birds.

IV

Intelligence resides in the sparrow's beak
And the seat of the will is the wing of the wasp
I am here I am there and my mind is in the middle
I hold in my hands the knob of the door of sleep
I stand on my feet on the rock of the principle
And my eyes are on top of my head
They see all that happens in the sky
The horse that bears his master in his mouth
And is ridden by the girl with red plush breasts
My ears grow out of my feet
And they hear all the sounds underground
The ringing of bells in the caves
And the whisper of wandering roots
The intellect resides in the mineral's neck
And the seat of the soul is the mouth of the stone

Which is why the earth's veins are so stopped up with sand
And the sea is so full of green flame
For the earth is a kiss on the mouth of the sky
And the sky is a fan in the hand of the sun

V

This is my world this is your realm of clay
Our dreams have all come true
The ash of sleep is deeper than dust on the stairs
Of this mine-shaft brimmed with gold
The sunken garden of a fugitive
Cold with black rain that stains the soil like ink
Enigma like a skull with petrol eyes
A sprouting head of plumes of silver grass
That haunts the sanded paths
The booming caves are full of birds
With silken wings and beaks of solid stone
Who pass the time away
With burning feathers from their tails
In the flaming waterfall
This is my world this is your garden gate
Our vistas stretch a thousand leagues from here
As far as forests full of moving trees
As far as fingers holding tigers' skins
As far as bushes on the window-sill
As far as castles with unlicensed towers
As far as caskets full of human hair
As far as clouds on fire and dying swans
On lakes that swallow beds as fast as tigers swallow hands.

VI

On the sidewalks of New York
There are women who pass to and fro with napkins wrapped round
    their heads
So that no one can see their eyes
And machines lean out of the windows to record the number of
    their footsteps
A record is made of the sound of falling coins
That cover the streets with silver and cause fruit to ripen in bowls
And the lift-boys chant:
The sea comes once too often up the street
And the wind goes once too seldom down the sky,
And their song goes on till morning

When the inhabitants put logs outside their doors
For the children to make fires in all the gutters
Which awakens the town to the sound of derailed trains
While baskets of boot-buttons light up the distant hills.

VII

Undoubtedly the sun has burnt his hands
Undoubtedly the corn has grown too high
And when this is done
The first-class trains will stop running every afternoon at five
    o'clock
And the passengers next morning will alight
In a ditch of frozen milk
Their thoughts will return with regret to their twice-locked trunks
Full of borrowed dresses and discarded wedding-rings
They will groan with dismay at the thought of the coming day
Full of empty bags and crumbs of stalest bread
From house to house the frost will spread its warnings
And weathercocks fall from the roofs.

VIII

The needle glitters inch by inch
And the sound of its stitches reaches the sea
Where bombs explode in every other wave
And the beaches are paler than curd
I return there every other night
Wearing the same clothes, breathing the same air
And the weasels only laugh at me but it is not my fault
I can hardly help it if the lines of the meridian resemble fish
That fly away
To where the heat softens the equator
With hair growing out of its ears
And birds' nests in its hair to keep the rain off
The rain that whispers in decrepit castles
Great clots of clay and the effigies falling to dust
Preserve us from the singing towers
And the chapter which turns the page of its own accord
For fear of reading its own history there.

THE SUPPOSED BEING

Supposing the mouth
The hard lips crowned with bright flowers
A bursting foam of petals
And each gold stamen an anxious arrow
As each firm finger a signal
Pointing to fire and water's junction
Whose furious fumes would stifle the passers-by
With their startled eyes
With their nervous hands and faces
Whose language is black whose language has
Never been ours.

Supposing the eyes
Luscious in lashes and deep stained with sleep
The eyes in the forehead like pools in the rocks
And the turbulent sea approaching
Shivering ravenous venomous scarred
By the sharp-taloned claws of its waves
As eyes by their ravaging lids
As their lids by the richly veined hands
That are burnt by the light of the sun
And the stones are on fire
And the pupils of eyes are glazed by the
Heat of their flames.

Supposing the hands
With their nails and their delicate bones
Like the frail limbs of birds
And their tips like the pink tips of buds
That probe the cold curious air
And discover the blood neath the skin
And the surface of stones.

Supposing the breasts
Like shells on the oceanless shore
At the end of the world
Like furious thrusts of a single knife
Like bread to be broken by hands
Supposing the breasts still untouched by desires
Still unsuckled by thirsts
And motionless still

Breasts violently still and enisled in the
Night and afraid both of love and of death.

Supposing the sex
A cruelty and dead in the thighs
A gaping and blackness—a charred
Trace of feverish flames
The sex like an X
As the sign and the imprint of all that has gone before
As a torch
To enlighten the forest of gloom and the
Mountains of unattained night.

And supposing the being entire
The tangible body standing
The visible limbs existing
And moving across the daylight
Or motionless in the darkness
A stone on the torrent's bed
Or a torrent above the stones—
And at last

Such a being escapes from the sight of my visible eyes
From the touch of my tangible hand
For she only exists
Where all contradictions exist
Where darkness is light and the real is unreal and the
World is a dream in a dream.

THREE VERBAL OBJECTS
In Memory of Humphrey Jennings

I

The poet is dead; and it is in the people that we must seek to find what
remains of the mysterious radiation of his soul:—birthpangs of a series
of images stretching away into infinity; crystallization of the
movements of impulsion and repulsion; from the hermit's cave to the
broken shell of the great roc, a trail of bones and other fragments.

In the centre of the arc is fixed a hunter's bow and arrow, festooned
with deadly flowers. This is the node of animal magnetism and of all
dreams of hate and fear. The people have secretly proclaimed their
love for those who haunt them.

Over the marshes, in the summer air, there hang invisible monsoons which, if the human eye could register them, would have the form of funnels. Into their mouths pass the warm breath of sinking creatures and the emanations of defeated warriors, whose shields and armour glitter strangely in the green light of the setting sun. Some representative of a distant tribe is seated passive on the bank, occasionally beating an idle note upon his drum. There will be no more thunder for at least another month.

The people love the warrior; and even as he lies sinking in the marsh, they deck his image with a thousand lethal flowers. They cannot see his wounds.

To the warrior, war; to the lover, love. And the lower species also shall give instances of their passion: in the twilight crevices, silent bubbles, swelling and deflating like the lungs. Smoke rises out of the eyes, distorting the labyrinthine perspective. This is sleep. Its oscillations only serve to aggravate the decay of the outer ramparts. There, our projected bodies parade themselves, clad in all the amazing appearance of the illusions which, without knowing it, we can entertain about them in the night.

Violent is the falsehood with which we have clothed our desires. To eat, to kill and to make love. Magic, the clotted valve, intoxication, cold invading the pores, the syren-call of giddiness falling from North to South. The ocean does not cease to lacerate the shore; nor the blood to circulate through the channels of the brain.

II

It is well-nigh impossible to describe in words the natural beauties of this country. The hills are bathed in a glow of the most subliminal tranquility, like that which is given out by the innocent eyes of children, milky and diffuse. The shadows cast by the further ranges eat into the plain like acid. There are only a few houses round the edge of the lakes, dwelling-places of fakirs and water-diviners, untroubled spirits who appear on their thresholds only at evening, when the sun throws an additional lustre upon the bismuth grottoes which adorn the shores. Who would not envy them, who pass their days in the ecstatic contemplation of the death of Time? To the North, there stand the remains of one or two deserted villages. These were once inhabited by an outlandish race, wearing skins and communicating with one another in a speech most closely resembling that of birds, shrill yet guttural. Their wells ran dry, or became salt, and they migrated, we know not where. In their abandoned huts, which were hewn from volcanic stone, a few pots and other utensils have been discovered, bearing curious ornamentations which are supposed to illustrate the myths of the lost tribe.

Foremost among these legendary representations is the Wheel. Sometimes this fantasy is expressed as a chain of limbs of animals and of men entwined. Some vessels, again, are covered with what appear to be crudely drawn bands of flowers. Certain monoliths, also, which have recently been discovered among the petrified Western forests, where the ground is frequently shaken by seismic tremors, are ornamented, totem-like, by circular constellations of five eyes.

The phenomenon of the wheel of eyes is said to have been frequently observed in this part of the country by watchers on the hills at dawn. The last occasion on which it was reported to have been seen was when a band of scientifically-minded explorers were making their way into the extreme fastnesses of the nether mountains, not so many years ago. They were emerging from their tents, at about five o'clock in the morning, when their guides drew their attention to a curious patch of light in the sky just above their heads. A few moments later, it became quite clear that this light was being given out not, as they had at first imagined, by a cluster of stars, but by five enormous and distinctly outlined eyes, which hovered gravely, motionless and without blinking in the sky for about ten minutes, holding the spectators spell-bound with silent awe and wonder, and then faded away like cloud.

The phenomenon was accompanied by a distant grinding, ringing sound. It is supposed that this mirage, or optical illusion, is due to the peculiar reflective properties of the mica rocks with which the region is encumbered; and that it was this appearance which originally gave rise, in association with the other and more obvious symbols of perpetual recurrence, to the legend of the Wheel.

III

Vast expanses of devastated territory, jagged skyline, wooden scaffoldings 140 foot high and blazing like giant torches—young women and little old children lying murdered in disordered heaps— abandoned gun-carriages, drifts of snow lying melting in the sun here and there among the ruins . . .

Everything was in order. Our leader called a halt. He turned his face towards us, away from the shattered landscape, and we saw that he was smiling through his tears. When the last trace of the old world is cleared away, comrades, he cried, we shall build our city here.

And one of our number planted the standard on top of a hillock of refuse. We set to work with diligence in the fresh morning smell. Everything is in order, we repeated to ourselves, looking up now and then to observe the destruction of a last altar, or a prison wall.

From the tower of a quietly blazing mansion whirled a flock of doves, and the smell of their half-scorched feathers became confused with the scent of the countless damp and trampled plants that lay a-rotting on the terraces. And the sky flung a column of wind like a wide-flung scarf into the distance, where the earth was turning on its never-ending hinge.

PHANTASMAGORIA
For Margaret W.

The wind has stopped at last
in that little black town on the edge of a violet sea
where a man in an upstairs-room of the empty house
which stands overlooking the yard of the Sodium Works
is sitting blindfold on the draughty floor
trying to hear the feeble groans of the North Pole inside his skull
and thinking of the iron teeth of Death
thinking of the rusty police-whistle chained to so many necks
of the last Act of *Faust*
of the cherry-coloured gown his mistress wore on that fatal night
   when she lost her head so irretrievably while sailing in a gondola
and of the incomparably curvilinear and seductive effect to be
   obtained
by writing one's name in water
with the white of one's own glass eye . . .
In this poor blackened town on the edge of a violet sea
the wind has left stray locks of hair behind
in almost every street—
locks which appear like loosely-knotted strands of twilight sleep
or fragments of Opal-tree bark
preserved in wine
and left all night to dry upon the steps of a Russian church . . .
These scattered tresses make the passers-by turn pale
then hurry home to disinfect their wells
They glitter faintly like the dust of poisoned stars
and hypnotize the gaze of the last birds still to remain
in that seaside town as black as a burnt cake

where the dead are sitting propped-up in the windows robed in
    flags
of all the nations—where the homeless night
is kept awake by Autumn's chill aurora in the sky
and silence lolls like smoke along the disused harbour-quays . . .
And in this little town like a charred bun beside a sea
which stains its shores with blackberry-juice ink
the crowds continue playing their quaint melancholy games
in street and market-place although dense clouds of smoke
are pouring from the windows of the Luxury Hotel
in which the foreign guest in Room 13
swathed in red bandages from head to foot
lies thinking of the monkey's-paw of Death
thinking of the frozen music in the eyes of statues
of the brutal naked beauty of a surgical machine
of his father's raincoat gleaming in the twilight long ago
and of the fungus growing on the tree-trunk of Desire . . .
In that charcoal-black town on the edge of a vein-coloured sea
where shadow smoulders in the cave-like shops
and copper bells toll slowly all day long
the wheels of a great lacquered Rolls-Royce car
left lying in the middle of the main street upside-down
are to be seen months later still continuing to spin
in the tensely sensational glare of the naphtha torch
left burning there by the authorities to mark the fatal spot—
continuing still to spin like a soul in pain
like a tin-plate sent whirling out without a word through the
    window-bars of a condemned man's cell
or like the breasts of Destiny revolving night and day . . .
And now that the day's white wind has stopped at last
the hoofs of dusk go trampling through the hollow clouds on high
from beneath their rocks the scorpions of the darkness soon creep
    out
and faintly in the distance on all sides is to be heard
the dread hyena-laughter of the prehistoric Night . . .
Meanwhile through narrow twilit streets flock jostling throngs of
    masks—
red oblong leather faces stuck with clusters of tiny shells
faces of cheese with green protruding fangs
faces like pillows wet with tears and moulting feathers through the
    torn holes of their eyes
and snarling hairy faces like the hindquarters of apes
and sickly faces weak as greasy smudges left by flies
and hungry faces gaping like raw muddy graves in Spring . . .

The thoroughfares of Evening swarm with rapid shifting scenes
and everywhere the lamps of lust and terror thrust their beams
to scour the countless cage-like haunts of men with scorching light
while waves of sound roll out across the rooftops overhead—
waves swollen with dreamy cries and rumbling words
with the last thick sobs of harlots stabbed to death
and with that unbearably heart-rending melody which the blind old
    men who live alone in freezing garrets are forever playing to
    themselves upon their broken violins . . .
See! here is a ring of dancers round a blazing marriage-bed
and here is a bunch of bearded dwarfs dangling chained by their
    heels from the top of a convent-wall
and here are the bones of a Saint which calmly float
upon the silken surface of a swimming-pool hewn from the heart of
    an amethyst-rock
in a glass-panelled coffin of cork lit-up inside on the stroke of
    midnight by a magnesium-flare . . .
Here is the Theatre standing open to the sky
in which dead flowers and moonlight perform ballets once an hour
and there the Children's Home stands on the hill behind the town
where hidden in steep gardens among shadows and blue shrubs
an orphan whose huge head lolls like a glass-eyed hirsute globe
squats weeping in the dew-chilled herb of dreams
and thrusting the blade of his pen-knife ever deeper into his thigh
And here is the swift silhouette of a sphinx on a screen in the sky
Here is the abandoned saw-mill with its broken windows' haggard
    gaze
and see! here the pair of superb nocturnal swans
each of which has been saddled with a mirror and firmly trussed to
    the back of a mule
and the mules stationed as sentries on either side the harbour's
    mouth
where every now and then they are washed gently from side to side
    by the changing tide . . .
And here among the dunes are strewn the battered hulks of wrecks
which ere the hour is far advanced abruptly rise into the air
and like a furtive school of whales go lunging inland through the
    night
to make their clumsy nests on the most lofty towers and domes;
while here upon the beach is the vast ballroom with invisible glass
    walls
across the luminous floor of which a hundred pairs of invisible
    slippers are picking their way among numberless pools of invisible
    blood . . .

68

And O how pungent is the firedamp's musty fragrance in the hollow
of each wave
that falls on the shore by that small black-eyed town on the edge of
a heliotrope sea
where a man in a brilliantly illumined subterranean padded-cell
concealed at a depth of about 69 feet below the level of the
ground—
(a man wearing a mask designed to resemble the head of a
Paradise-bird
with a diamond-encrusted beak of solid gold
and clad in a sky-blue satin tunic across the front of which are
embroidered in silver thread
the words SPITTOON—OSMOSIS—SINGAPORE)—
sits swinging regularly to and fro upon a platinum trapeze
and thinking of the iridescent and immobile nipples of Death
thinking of the vivid short-lived blossoms which are seen to sprout
occasionally from the mouths of pregnant women
of how the midnight-sun drapes the landscapes of Arabia with
invertebrate question-marks like plumes snatched from an ailing
eagle's tail
of the colourless abyss of idle days
of Mary calling home the cattle across the sands of Dee
and of the end of Summer with its interminable showers of salt and
of soot . . .
But now that the great water-spouts of midnight have subsided out
at sea
and that those barbaric cortèges of clouds swaying dangerously from
side to side across the steeps of heaven
like sodden hayricks in a sudden storm
have finally all vanished one by one into the fuming workhouse-
chimneys of the East—
now that the cavernous yawn of the lonely female Titan lying
sleeping on the softly gleaming sands
has at last swallowed-up every starfish in sight—
the livid wind once more begins to lift,
stealthily weaving its fine-spun shawls in writhing swathes around
the radius of that small black seaside town
through which by now down each long soundless street
swarms of somnambulistic barefoot children creep
by slow degrees, still sealed by spell of dream,
towards where soon the spume-besilvered waves shall shine and
seethe
as a new Sun soars like song out of the silence of the sea.

IV

HÖLDERLIN'S MADNESS

(1937–1938)

HÖLDERLIN'S MADNESS

FIGURE IN A LANDSCAPE

The verdant valleys full of rivers
Sang a fresh song to the thirsty hills.
The rivers sang:
'Our mother is the Night, into the Day we flow. The mills
Which toil our waters have no thirst. We flow
Like light.'

       And the great birds
Which dwell among the rocks, flew down
Into the dales to drink, and their dark wings
Threw flying shades across the pastures green.

At dawn the rivers flowed into the sea.
The mountain birds
Rose out of sleep like a winged cloud, a single fleet
And flew into a newly-risen sun.

—Anger of the sun: the deadly blood-red rays which strike oblique
Through olive branches on the slopes and kill the kine.
—Tears of the sun: the summer evening rains which hang grey veils
Between the earth and sky, and soak the corn, and brim the lakes.
—Dream of the sun: the mists which swim down from the icy
   heights
And hide the gods who wander on the mountainsides at noon.

The sun was anguished, and the sea
Threw up its crested arms and cried aloud out of the depths;
And the white horses of the waves raced the black horses of the
   clouds;
The rocky peaks clawed at the sky like gnarled imploring hands;
And the black cypresses strained upwards like the sex of a hanged
   man.

*

Across the agonizing land there fled
Among the landscape's limbs (the limbs
Of a vast denuded body torn and vanquished from within)
The chaste white road,
Prolonged into the distance like a plaint.

Between the opposition of the night and day
Between the opposition of the earth and sky
Between the opposition of the sea and land
Between the opposition of the landscape and the road
A traveller came
                    Whose only nudity his armour was
Against the whirlwind and the weapon, the undoing wound,

And met himself half-way.

Spectre as white as salt in the crude light of the sky
Spectre confronted by flesh, the present and past
Meet timelessly upon the endless road,
Merge timelessly in time and pass away,
Dreamed face away from stricken face into the bourn
Of the unborn, and the real face of age into the fastnesses of death.

Infinitely small among the infinitely huge,
Drunk with the rising fluids of his breast, his boiling heart,
Exposed and naked as the skeleton—upon the knees
Like some tormented desert saint—he flung
The last curse of regret against Omnipotence.
And the lightning struck his face.

*

After the blow, the bruised earth blooms again,
The storm-wrack, wrack of the cloudy sea
Dissolve, the rocks relax,
As the pallid phallus sinks in the clear dawn
Of a new day, and the wild eyes melt and close,
And the eye of the sun is no more blind—

Clear milk of love, O lave the devastated vale,
And peace of high-noon, soothe the traveller's pain
Whose hands still grope and clutch, whose head
Thrown back entreats the guerison
And music of your light!

The valley rivers irrigate the land, the mills
Revolve, the hills are fecund with the cypress and the vine,
And the great eagles guard the mountain heights.
Above the peaks in mystery there sit
The Presences, the Unseen in the sky,
Inscrutable, whose influences like rays
Descend upon him, pass through and again
Like golden bees the hive of his lost head.

ORPHEUS IN THE UNDERWORLD

Curtains of rock
And tears of stone,
Wet leaves in a high crevice of the sky:
From side to side the draperies
Drawn back by rigid hands.

And he came carrying the shattered lyre,
And wearing the blue robes of a king,
And looking through eyes like holes torn in a screen;
And the distant sea was faintly heard,
From time to time, in the suddenly rising wind,
Like broken song.

Out of his sleep, from time to time,
From between half-open lips,
Escaped the bewildered words which try to tell
The tale of his bright night
And his wing-shadowed day
The soaring flights of thought beneath the sun
Above the islands of the seas
And all the deserts, all the pastures, all the plains
Of the distracting foreign land.

He sleeps with the broken lyre between his hands,
And round his slumber are drawn back
The rigid draperies, the tears and wet leaves,
Cold curtains of rock concealing the bottomless sky.

TENEBRAE

Brown darkness on the gazing face
In the cavern of candlelight reflects
The passing of the immaterial world in the deep eyes.

The granite organ in the crypt
Resounds with rising thunder through the blood,
With daylight song, unearthly song that floods
The brain with bursting suns:
Yet it is night.

It is the endless night, whose every star
Is in the spirit like the snow of dawn,
Whose meteors are the brilliance of summer,
And whose wind and rain
Are all the halcyon freshness of the valley rivers,
Where the swans,
White, white in the light of dream,
Still dip their heads.

Clear night!
He has no need of candles who can see
A longer, more celestial day than ours.

EPILOGUE

This severed artery
The sand-obliterated face
Amazed eyes high above catastrophe
Distributed—Is this the man's remains
Who walked the lap of lands, and sang?

Explosions of every dimension
Directions run away
Towards the sun
The bitter sunset, or
Who knows, where all things rise and fall,
Revolve, and meet themselves again?

This is the man of matted hair
And music, whom a wanderer
Had scented a long way off, by reason of
The salt blood in his heart,
The black sun in his blood,
The gestures of his skeleton, simplicity
Of white bones worn away
Like rock by milk of love.

Dissolve and meet themselves again
All things; the sandy artery
The severed head
Limbs strewn across the rocks
Like broken boats:
So shall their widespread body rise
And march, and marching sing.

V
THE CONSPIRATORS
(1939)

THE CONSPIRATORS

PRELUDE TO AN UNFINISHED NARRATIVE

Here is the Capital.
                   'Observe
How like a microscopic slide whose glass arena holds
Spectacular combat of schizomycetes, these grappling streets
Elucidate with their contrasting quarters the disease
Disintegrating all these fated lives. Lives of the refuse-heap, the
Slum, the rusty dump, packed in a fouled dilapidated bed,
Running with sores for years like washerless taps. The lives
Of eremites, black-coated, in their desert no-man's-land
Of tidy, sterile, separate brick cells, pitched just half-way
Between the catacombs of want and the gilt mansions of big pots.
Lives of the latter, lush as scum on standing water, limply led
Through periods of alternating boredom, frantic spending and
Bewilderment, by an unhappy little race of monsters
Caricatured by Grosz, staring with fascinated eyes
At their own image on the cruel plate-glass their diamonds
Cannot break.'

           Thus speaks the voice of the didactic guide
In the intelligentsia-tourist charabanc. But let's remove
Clinical spectacles, look round with naïve gaze: Here slide
The undramatic trams crammed with normality, shop-windows
    greet
The morning housewife with their pyramid displays, and children
Chase callous hoops among thin legs along the curb. But O
The glamour of the metropolitan sunlight, coffee smells,
Striped awnings, the bright water-dust of fountains! O the
Pigeons, scattering foam of wings! 'Call me a taxi!' 'Midday news!
New Cabinet Formed. All Racing Form.' '. . . We'll meet you in
The Park.' 'The Ritz for cocktails . . .'

               Surface appeal conceals
With fragrant clouds the city's noisesome heart, as the façades
Of these white buildings flecked with flags and
Flowering window-boxes can divert a strolling eye
With the irrelevance of statues' nudity, so hide
The dramas in their bowels: the Senate House and the
State Hospital. The Institute of Science, where the famous
Flambow lectures.

                    This same afternoon
The National Socialist minority in the Senate rose

Up in a body, shouting: 'Treacherous reds! If we
Resign we shall return in triumph, set this
House in order in the people's name!' Their barking
Met with smiles from liberal benches. When a telephone
Called for a left-wing member, he returned with a white
Face: 'Max Kleinborg, Jewish leader, died in hospital
An hour ago. Mysterious injuries. Unknown
Assailant.' No one smiled again. At the same hour
In a packed lecture-theatre at the Institute
Flambow declared: 'Our highest of ideals
Is to maintain and serve the freedom of research that we
Have won. I do appeal to every student here
Never to sacrifice the human interest to any such demands
As may be made on us by an exterior cause. When I was asked
To aid the government by giving up my time
To the discovery of new poison-gas, explosives and death-rays,
I categorically refused.' Bursting applause
Completed his last words; but from the shadows at the back of
The long hall, an angry cry: 'The Fatherland
Comes before all! Flambow, beware!'
                                    Clapping of hands,
Raised voices. Heard down the corridor. Third floor,
First on the right, door 17: the Faculty of
Sociological Studies, where reports from the anonymous
Observer are received and filed (under the supervision of
Jules Hartmann, son of Flambow's greatest friend). Today
A busy afternoon. Piles of thick sheaves whose contents
'Plot on a graph that tortuous nervy line, the mass's
Changing life.'
                    Chosen at random:

                    'Rose
At half-past five. Argued at breakfast with the wife over the
Pending strike. Quit house at six. Rode through the rain
To work. Outside the gates a Grey-shirt stood distributing
His party's pamphlets (paid for by funds subscribed to by
Our boss). One of my mates went up to him and wrenched
The bundle from his hands. Bit of a scuffle. Later saw,
Lounging at lunch-hour, leaflets in the mud.'

                         '. . . to tea
With a professor and his wife in Tower Street. A Madame D.,
A well-dressed, cultured-looking woman, said she thought
That life was meaningless. Professor shrugged. The conversation
   turned
To table-turning and astrology.'

                         'One of the girls
In our department came to work today with a framed
Photo of "the Leader", as she calls him, which she stuck
Over her desk.'

                         'After the children had gone back
To school, I went up to lie down, as every day, but could
Not rest because of worry over what last night my
Husband said about his job, how he might lose it soon.'

'First came the standard-bearers clothed in tiger-skins, and then
A squad of troopers at salute. The band struck up a fanfare, and
Through curtains stepped the hero of the evening. The crowd's
   cheers
Were deafening. One woman fainted and was carried out. At last
He raised his hand. "My people!" he began, and then I heard
A man behind me say "Not yet, thank God!" At once
He disappeared beneath a dozen blows.'

                         'As I
Was leaving the Exchange, a fellow said to me that if
The NS party were in power there'd be no unemployment
Benefit. He'd rather die, he said. He used to be
On the same shift with me. We strolled to the disused pithead,
A car was standing there, drove off as we came by.'

'The street was full of people and I saw a van
Loaded with special police arrive, but they were not
Able to make the rioters disperse. Then someone shouted:
"Let them have it, they're his bloody guards!" That started
It. I noticed that a clock said half-past ten. Then I was knocked
Down by the baton charge.'

                              So would a seismograph describe
Its dire parabolas. The scattered records utter all the same
Act, act, to Hartmann's ear. How can one hear them, impotently
    tied
By scientific objectivity, he urgently inquires. The will of one
To climb upon the roof-top of the Institute, launch a premonitory
    cry
Like meteoric words of sky-sign smoke across the town
To hang there hugely inescapable, for all to see, subsides
A disillusioned wave in him. 'What can I do
But urge my Father to persuade the leading men of the
Executive to issue an immediate appeal
For unity, to act, to act, throughout
The workers' movement. Soon will be too late.'

But evening takes its coat down from the peg,
Portals are closing, private lives resume
Their homechat-crossword puzzles and the knitting of
Protective woollies: armies evacuate
The daily battlefield, and clad in mufti roam
Through park, arcade and alley whistling gay
Or wistful tunes, not marching-songs. And though
The hour's as heavy as a pear tense on its bough
Awaiting a mere puff to make it drop, a ripened fruit
Swollen with change and danger like a bomb, only a few –
The soothsayer, the seer, the rebel poet – see it there
Suspended in the sunset, ominous.

                              O evening flares
Placard this town and country with perfervid colours! O
Remember, when the coming night is thick and weak the pulse
Of hope and under cover of the dark your freedom's last
Defenders have deserted or been shot, remember this
Dazzling finale! Music in the parks and lights beneath
The trees, where the loudspeakers not yet blare
With only race-hysteria; crimson lakes
Poured out across the heavens that do not as yet
Reflect a nation's blood; on outward roads
Car-fleets that are not freighted yet with loads
Of refugees. In floodlit sapphire pools
Still swim the golden poignant limbs of youths
Unregimented, girls for whom kisses do not seal
A cannon-fodder contract. On the greens
Children play games untouched by creed or badge,

Not yet corrupted by the partisan's
Crude flag.

                                    Flambow, returning late across
The City Gardens from laboratory, heard
Their mothers call them home, and sighed, and sought
Not to imagine how those voices might ascend
One day in agonized crescendo, how the blooms
In the neat beds might be replaced by red
Flowers of carnage, and that placid lawn
Be suddenly transformed into a desert waste
Littered with bones and stony fragments. 'Peace
Is our most precious ally to defend: my work
Is unrelenting undestructive war against war's works
And evil allies.' Overhead the air
Condensed the overtones of dusk, and at his door
He turned awhile to gaze up at a star.

The clock says Night. Now the conspirators
Assemble: now in the centre of the town within
The wooden horse of the Grey House the shirted band
Prepare their fatal coup round shaded lamps
Which drop white circles on their charts and black-
Marked lists. Passwords, salutes and codes observe
Their midnight ritual. Assassinations brew
In shady cafés: while in the frank glare
Of chandeliers the Leader drinks champagne,
The guest tonight of patriotic heads
Of certain industries, not slow to recognize
Their Saviour. Trusting to dreams less well-insured
The people plunge into the fogs of sleep
Through which they drift towards tomorrow's rocks.

VI

MISERERE AND OTHER POEMS
(1937–1942)

MISERERE

'Le désespoir a des ailes
L'amour a pour aile nacré
Le désespoir
Les sociétés peuvent changer.'
PIERRE JEAN JOUVE

TENEBRAE

'It is finished.' The last nail
Has consummated the inhuman pattern, and the veil
Is torn. God's wounds are numbered.
All is now withdrawn: void yawns
The rock-hewn tomb. There is no more
Regeneration in the stricken sun,
The hope of faith no more,
No height no depth no sign
And no more history.

Thus may it be: and worse.
And may we know Thy perfect darkness.
And may we into Hell descend with Thee.

PIETÀ

Stark in the pasture on the skull-shaped hill,
In swollen aura of disaster shrunken and
Unsheltered by the ruin of the sky,
Intensely concentrated in themselves the banded
Saints abandoned kneel.

And under the unburdened tree
Great in their midst, the rigid folds
Of a blue cloak upholding as a text
Her grief-scrawled face for the ensuing world to read,
The Mother, whose dead Son's dear head
Weighs like a precious blood-encrusted stone
On her unfathomable breast:

Holds Him God has forsaken, Word made flesh
Made ransom, to the slow smoulder of her heart
Till the catharsis of the race shall be complete.

89

DE PROFUNDIS

Out of these depths:

Where footsteps wander in the marsh of death and an
Intense infernal glare is on our faces facing down:

Out of these depths, what shamefaced cry
Half choked in the dry throat, as though a stone
Were our confounded tongue, can ever rise:
Because the mind has been struck blind
And may no more conceive
Thy Throne . . .

Because the depths
Are clear with only death's
Marsh-light, because the rock of grief
Is clearly too extreme for us to breach:
Deepen our depths,

And aid our unbelief.

KYRIE

Is man's destructive lust insatiable? There is
Grief in the blow that shatters the innocent face.
Pain blots out clearer sense. And pleasure suffers
The trial thrust of death in even the bride's embrace.

The black catastrophe that can lay waste our worlds
May be unconsciously desired. Fear masks our face;
And tears as warm and cruelly wrung as blood
Are tumbling even in the mouth of our grimace.

How can our hope ring true? Fatality of guilt
And complicated anguish confounds time and place;
While from the tottering ancestral house an angry voice
Resounds in prophecy. Grant us extraordinary grace,

O spirit hidden in the dark in us and deep,
And bring to light the dream out of our sleep.

LACHRYMAE

Slow are the years of light:
                                    and more immense
Than the imagination. And the years return
Until the Unity is filled. And heavy are
The lengths of Time with the slow weight of tears.
Since Thou didst weep, on a remote hill-side
Beneath the olive-trees, fires of unnumbered stars
Have burnt the years away, until we see them now:
Since Thou didst weep, as many tears
Have flowed like hourglass sand.
Thy tears were all.
And when our secret face
Is blind because of the mysterious
Surging of tears wrung by our most profound
Presentiment of evil in man's fate, our cruellest wounds
Become Thy stigmata. They are Thy tears which fall.

EX NIHILO

Here am I now cast down
Beneath the black glare of a netherworld's
Dead suns, dust in my mouth, among
Dun tiers no tears refresh: am cast
Down by a lofty hand,

Hand that I love! Lord Light,
How dark is Thy arm's will and ironlike
Thy ruler's finger that has sent me here!
Far from Thy face I nothing understand,
But kiss the Hand that has consigned

Me to these latter years where I must learn
The revelation of despair, and find
Among the debris of all certainties
The hardest stone on which to found
Altar and shelter for Eternity.

SANCTUS

Incomprehensible—
O Master—fate and mystery
And message and long promised
Revelation! Murmur of the leaves
Of life's prolific tree in the dark haze
Of Midsummer: and inspiration of the blood
In the ecstatic secret bed: and bare
Inscription on a prison wall, 'For thou shalt persevere
In thine identity . . .': a momentary glimpsed
Escape into the golden dance of dust
Beyond the window. These are all.

Uncomprehending. But to understand
Is to endure, withstand the withering blight
Of winter night's long desperation, war,
Confusion, till at the dense core
Of this existence all the spirit's force
Becomes acceptance of blind eyes
To see no more. Then they may see at last;
And all they see their vision sanctifies.

ECCE HOMO

Whose is this horrifying face,
This putrid flesh, discoloured, flayed,
Fed on by flies, scorched by the sun?
Whose are these hollow red-filmed eyes
And thorn-spiked head and spear-stuck side?
Behold the Man: He is Man's Son.

Forget the legend, tear the decent veil
That cowardice or interest devised
To make their mortal enemy a friend,
To hide the bitter truth all His wounds tell,
Lest the great scandal be no more disguised:
He is in agony till the world's end,

And we must never sleep during that time!
He is suspended on the cross-tree now
And we are onlookers at the crime,
Callous contemporaries of the slow

Torture of God. Here is the hill
Made ghastly by His spattered blood

Whereon He hangs and suffers still:
See, the centurions wear riding-boots,
Black shirts and badges and peaked caps,
Greet one another with raised-arm salutes;
They have cold eyes, unsmiling lips;
Yet these His brothers know not what they do.

And on his either side hang dead
A labourer and a factory hand,
Or one is maybe a lynched Jew
And one a Negro or a Red,
Coolie or Ethiopian, Irishman,
Spaniard or German democrat.

Behind His lolling head the sky
Glares like a fiery cataract
Red with the murders of two thousand years
Committed in His name and by
Crusaders, Christian warriors
Defending faith and property.

Amid the plain beneath His transfixed hands,
Exuding darkness as indelible
As guilty stains, fanned by funereal
And lurid airs, besieged by drifting sands
And clefted landslides our about-to-be
Bombed and abandoned cities stand.

He who wept for Jerusalem
Now sees His prophecy extend
Across the greatest cities of the world,
A guilty panic reason cannot stem
Rising to raze them all as He foretold;
And He must watch this drama to the end.

Though often named, He is unknown
To the dark kingdoms at His feet
Where everything disparages His words,
And each man bears the common guilt alone
And goes blindfolded to his fate,
And fear and greed are sovereign lords.

The turning point of history
Must come. Yet the complacent and the proud
And who exploit and kill, may be denied—
Christ of Revolution and of Poetry—
The resurrection and the life
Wrought by your spirit's blood.

Involved in their own sophistry
The black priest and the upright man
Faced by subversive truth shall be struck dumb,
Christ of Revolution and of Poetry,
While the rejected and condemned become
Agents of the divine.

Not from a monstrance silver-wrought
But from the tree of human pain
Redeem our sterile misery,
Christ of Revolution and of Poetry,
That man's long journey through the night
May not have been in vain.

METAPHYSICAL POEMS:

'Without cease and forever there is celebrated the
Mystery of the Open Tomb, the Resurrection of Osiris-Ra,
the Increated Light.'

The Book of the Dead

'Therefore it is said: And the deeper secret within the
secret: the land that is nowhere, that is the true home.'

The Secret of the Golden Flower

WORLD WITHOUT END

See how across the seas of azure milk
Transpire the changing tranquil cloudy forms
Which image us below. The other eyes
Profoundly sunken in us, brim
With such refractions and mysterious
Broken light-webs from the depths
Or inward heights.

         And without cease
The spirit's upward exhalation stirs

Susurrus and whirled currents of the central flame
Which burns relentlessly away
The lower body and the crystal skull
To carbon purity, and shines
Intense as daybreak down the rocky shafts
Into the world beyond.

INFERNO

One evening like the years that shut us in,
Roofed by dark-blooded and convulsive cloud,
Led onward by the scarlet and black flag
Of anger and despondency, my self:
My searcher and destroyer: wandering
Through unnamed streets of a great nameless town,
As in a syncope, sudden, absolute,
Was shown the Void that undermines the world:

For all that eye can claim is impotent—
Sky, solid brick of buildings, masks of flesh—
Against the splintering of that screen which shields
Man's puny consciousness from hell: over the edge
Of a thin inch's fraction lie in wait for him

      Bottomless depths of roaring emptiness.

LOWLAND

Heavy with rain and dense stagnating green
Of old trees guarding tombs these gardens
Sink in the dark and drown. The wet fields run
Together in the middle of the plain. And there are heard
Stampeding herds of horses and a cry,
More long and lamentable as the rains increase,
From out of the beyond.
                  O dionysian
Desire breaking that voice, released
By fear and torment, out of our lowland rear
A lofty, savage and enduring monument!

MOUNTAINS

Pure peaks thrust upward out of mines of energy
To scar the sky with symbols of ascent,
Out of an innermost catastrophe—
Schismatic shock and rupture of earth's core—
Were grimly born.
                              O elemental statuary
And rock-hewn monuments, whose shadow we
Lie low and wasting in, a prey to inner void:
Preach to us with great avalanches, tell
How new worlds surge from chaos to the light;
And starbound snowfields, fortify
With the stern silence of your white
Our weak hearts dulled by the intolerably loud
Commotion of this tragic century.

WINTER GARDEN

The season's anguish, crashing whirlwind, ice,
Have passed, and cleansed the trodden paths
That silent gardeners have strewn with ash.

The iron circles of the sky
Are worn away by tempest;
Yet in this garden there is no more strife:
The Winter's knife is buried in the earth.
Pure music is the cry that tears
The birdless branches in the wind.
No blossom is reborn. The blue
Stare of the pond is blind.

And no one sees
A restless stranger through the morning stray
Across the sodden lawn, whose eyes
Are tired of weeping, in whose breast
A savage sun consumes its hidden day.

THE WALL

At first my territory was a Wood:
Tanglewood, tattering tendrils, trees
Whose Grimm's-tale shadow terrified but made
A place to hide in: among traps and towers
The path I kept to had free right-of-way.

But centred later round an ambushed Well,
Reputed bottomless; and night and day
My gaze hung in the depths beneath the real
And sought the secret source of nothingness;
Until I tired of its Circean spell.

Returning to the narrow onward road
I find it leads me only to the Wall
Of Interdiction. But if my despair
Is strong enough, my spirit truly hard,
No wall shall break my will: To persevere.

THE FORTRESS

The socket-free lone visionary eye,
Soaring reflectively
Through regions sealed from macrocosmic light
By inner sky's impenetrable shell,
Often is able to descry:
Beyond the abdominal range's hairless hills
And lunar chasms of the porphyry
Mines; and beyond the forest whose each branch
Bears a lit candle, and the nine
Zigzagging paths which lead into the mind's
Most dangerous far reach; beyond
The calm lymphatic sea
Laving the wound of birth, and the
Red dunes of rot upon its further shore:

97

A heaving fortress built up like a breast,
Exposed like a huge breast high on its rock,
Streaming with milky brightness, the domed top
Wreathed in irradiant rainbow cloud.
                                        The shock
Of visions stuns the hovering eye, which cannot see
What caverns of deep blood those white walls hide,
Concealing ever rampant underneath
The dark chimera Death-in-life
Defending Life from death.

DICHTERSLEBEN

Lodged in a corner of his breast
Like a black hole torn by the loss
Of an ancestral treasure, like a thorn
Implanted ineradicably by his first
Sharp realization of the world, or like a cross
To which his life was to be nailed, he bore
Always the ache of an anxiety, a grief
Which nothing could explain, but which some nights
Would make him cry that he could fight no more.

Time ploughed its way through him; and change
Immersed him in disorder and decay.
Only the strange
Interior ray of the bleak flame
Which charred his heart's core could illuminate
The hidden unity of his life's theme.

He knew how the extremity of night
Can sterilize the final germ of faith;
Appearance crushed him with its steady weight;
Futility discoloured with its breath
His tragic vision. All his strength was spent
In holding to some sense from day to day . . .
Slowly he fell towards dismemberment.

Yet when he lay
At last exhausted under his stilled blood's
Thick cover and eyes' earth-stained lids,
The constant burden of his breast
(Long work of yeast) arose with joy
Into its first full freedom, metamorphosised, released.

I.M. BENJAMIN FONDANE
(1898–1944)

This is the osseous and uncertain desert
And valley of death's shadow, where the desired
Sweet spiritual spring is sought for
But unfound.
                        It is beyond
And far, and lost in the essential blue
Of space, among the rock and snow, the locked
Domain the instinct asks for. They who wait
Without the great thirst of despair are cursed;
And they who quench their thirst in death
Shall fall asleep among the mirages. But the
Inspired and the unchained and the endowed of desperate grace
Shall break through the last gate, by violence take
God's Kingdom, and attain the certain State.

MOZART: SURSUM CORDA
For Priaulx Rainier

Filters the sunlight from the knife-bright wind
And rarifies the rumour-burdened air,
The heart's receptive chalice in pure hands upheld
Towards the sostenuto of the sky

Supernal voices flood the ear of clay
And transpierce the dense skull: Reveal
The immaterial world concealed
By mortal deafness and the screen of sense,

World of transparency and last release
And world within the world. Beyond our speech
To tell what equinoxes of the infinite
The spirit ranges in its rare utmost flight.

CAVATINA

Now we must bear the final real
Convulsion of the breast, for the sublime
Relief of the catharsis; and the cruel
Clear grief; the dear redemption from the crime,
The sublimation of the evil dream.

Beneath, all is confused, dense and impure;
Extraordinary shiftings of a nameless mass
From plane to plane, then some obscure
Catastrophe:
                    The shattered Cross
High on its storm-lit hill, the searchlight eyes
Whose lines divide the black dome of the skies,
Are implicated; and the Universe of Death—
Gold, excrement and flesh, the spirit's malady,
A secret animal's hot breath . . .

Yet through disaster a faint melody
Insists; and the interior suffering like a silver wire
Enduring and resplendent, strongly plied
By genius' hands into the searching fire
At last emerges and is purified.

Its force like violins in pure lament
Persists, sending ascending stairs
Across the far wastes of the firmament
To carry starwards all our weight of tears.

ARTIST

Caught in a web, and crushed within a vice;
Watched by an Eye, but out of sight;
By a brand burnt, and wounded by
More keen a rustless blade than ever cut
This earth's black veins.—The voice
Of prophecy destroys the speaker. Bleak
As a scraped bone, the stony tablelands
On which he stands.—He cannot kill
The serpent of the blood: but his ghost shall.
Though armies of his enemy extend
In coiling ranks around his feet, still yet
Shall he transcend defeat, if his great wound
Be kept from healing.—ARTIST! hold that host
Once more at bay by offering your flesh
As sacrifice to the Void's mouth in your own breast!

INSURRECTION

Turbulence, uproar, echo of a War
Beyond our frontier: burning, blood and black
Impenetrable smoke that only blast
Of Archangelic trumpet could transpierce!
What savagery
And what inhuman crime,
What odour of hot iron, nocturnal flesh
Of sexual animal these uncouth cries invoke!
Till round the naked hill of rearing rock
With roaring torches suddenly emerge,
Shaking archaic instruments of strife,
Infernal armies sent us to avenge
The too-long-suffered tyranny and
Celebrated scandal of man's life!

LEGENDARY FRAGMENT

Below, in the dark midst, the opened thighs
Gave up their mystery. Myrrh, cassia
And spikenard obscurely emanated from
The inmost blackness. As from all around
There rose a heavy sighing and a troubled light:
Reverberated in the ears and eyes
And stunned the senses.
                              Thus the harlot queen
Was vanquished, while the outmost walls
Of that great town still echoed with her praise.

EVE

Profound the radiance issuing
From the all-inhaling mouth among
The blonde and stifling hair which falls
In heavy rivers from the high-crowned head,
While in the tension of her heat and light
The upward creeping blood whispers her name:
Insurgent, wounded and avenging one,
In whose black sex
Our ancient culpability like a pearl is set.

VENUS ANDROGYNE

With gaze impaired by heavy haze of sense
And sleep-dust, see: the blasphemy of flesh!
The breast is female, groin and fist are male,
But the red sphinx is hidden underneath the
Weed-rank hair: muscle and grain
Of man inextricably twined
With woman's beauty.

Stand up, thorn
Of double anguish born, and pierce
The gentle athlete flank, that fierce pain
May merge like honey with the spirit's blood,
Purging desire: with agony atone
For such abhorrent heresy of seed,
And weld twin contradictions in a single fire!

AMOR FATI

Beloved enemy, preparer of my death,
When there's no longer any garment left
To lessen the clenched impact of our limbs,
When there is mutual drought in our swift breath
And twin tongues struggle for the brim
Of swollen flood—an aching undertow
Sucking us inward—when the blood's
Lust has attained its whitest glow
And the convulsion comes in quickening gusts,
Speaking is fatal: Do not break
That vacuum out of which our silence speaks
Of its sad speechless fury to the star
Whose glitter scars
The heavy heaven under which we lie
And injure one another O incurably!

THE FAULT

To live, and to respire
And to aspire, to feel the fire
Urge upward through the mortal part and gain
Through burnt-out veins still higher!
But who has lived an hour
In the condemned condition of our blood
And not known how a wound like a black flower,
Exquisite and irreparable, can break
Apart in the immortal in us, or not felt
An intimation of the fault: to be alive!

THE DESCENT

Where everything sinks down,
Is petrified in its descent, as still as vast
Perspectives full of ragged mountain and
Black forest of mortality
And azure air,
Sink swollen slowly downward frozen tears.

All is reflected in that Angel's eye
Who sees beyond the inward depth
Into the glittering schist of the far floor.

Naked the beautiful remembered limbs
And downward clustering hung
And mirrored in the dark encircling floods;
Suspended like a wreath and tremulous
In the mysterious wind of their blind flight and fall:

Unnumbered wings: and Ah! voluminous
The cloudy chasm like a gasping mouth
From whence the last deep cry so throughly torn
Unseals the Sepulchre of holy rock.

THE OPEN TOMB

Vibrant with silence is the last sealed room
That fever-quickened breathing cannot break:
Magnetic silence and unshakably doomed breath
Hung like a screen of ice
Between the cavern and the closing eyes,
Between the last day and the final scene
Of death, unwitnessed save by one:

By Omega! the angel whose dark wind
Of wings and trumpet lips
Stirs with disruptive storm the clinging folds
Of stalagmatic foliage lachrymose
Hung from the lofty crypt, where endlessly
The phalanx passes, two by three, with all
The hypnotizing fall of stairs.

Their faces are unraised as yet from sleep;
The pace is slow, and down the steep descent
Their carried candles eddy like a stream;
While on each side, through window in the rock,
Beyond the tunnelled grottoes there are seen
Serene the sunless but how dazzling plains
Where like a sea resounds our open tomb.

THE PLUMMET HEART
In Memory of Hart Crane

Down, Hart, you fell down sound-
lessly, as though through shaft of lift,
leaving the roar of birth's wind-parted rift
around the topmost floor, no ground

beneath, no wreath of rock
to crown your exit from this crux;
and as you dropped through the restricted flux
of such duration as the clock

controls, on swift walls shone
in mirrors as you hurtled by
the scripture chiselled by your heart: until
the sea received you, azure antiphon
whose octave answer is the sky
where your wrecked smile drifts still.

THE THREE STARS

A PROPHECY

The night was Time:
The phases of the moon,
Dynamic influence, controller of the tides,
Its changing face and cycle of quick shades,
Were History, which seemed unending. Then
Occurred the prophesied and the to be
Recounted hour when the reflection ceased
To flow like unseen life-blood in between
The night's tenebral mirror and the lunar light,

Exchanging meaning. Anguish like a crack
Ran with its ruin from the fulfilled Past
Toward's the Future's emptiness; and *black*,
Invading all the prism, became absolute.

Black was the No-time at the heart
Of Time (the frameless mirror's back),
But still the Anguish shook
As though with memory and with anticipation: till
Its terror's trembling broke
By an unhoped-for miracle Negation's spell:
Death died and Birth was born with one great cry
And out of some uncharted spaceless sky
Into the new-born night three white stars fell.

And were suspended there a while for all
To see and understand (though none may tell
The inmost meaning of this Mystery).

The first star has a name which stands
For many names of all things that begin
And all first thoughts of undivided minds;
The second star
Is nameless and shines bleakly like the pain
Of an existence conscious only of its end,
And inarticulate, alone
And blind. Immeasurably far
Each from the other first and second spin;
Yet to us at this moment they appear
So close to one another that their rays
In one blurred conflagration intertwine:
So that the third seems born
Of their embracing: till the outer pair
Are separate seen again
Fixed in their true extremes; and in between
These two gleams' hemispheres, unseen
But shining everywhere
The third star balanced shall henceforward burn
Through all dark still to come, serene,
Ubiquitous, immaculately clear;
A magnet in the middle of the maze, to draw us on
Towards that Bethlehem beyond despair
Where from the womb of Nothing shall be born
A Son.

EPODE

Then
The great Face turned away in silence, veiled and slow,
Resigned and imperturbable: the brow
A grave dome drastic in its upthrust, and the eyes'
Unquenched blue fires of grief sealed and concealed
Beneath lids of irrevocable flint. It turned
Away; and as the shaft below began to slant
Towards its headlong fall into unknown
Futurity, the sacred Mouth enshrined
Like a sarcophagus within its midst revealed
During that moment's timeless flash
The wordless Meaning of the Whole
(Which may be spoken by no man)
Through the unearthly brilliance of its smile . . .

While the old world's last bonfires turned to ash.

PERSONAL POEMS:

SONNET: FROM MORN TO MOURNING

Morning. Full Chorus of the birds. A Sun
Of nascent ardour in the sapphire dome.
Now Memnon's massive kings with mouths of stone
Chant their aubade. Now down the valleys come
Innocent minstrels in whose unstained eyes
Vision unfolds vibrating like a flower:
Yggdrasil spreads above them; Jordan flows
About their feet; they hear the magic lyre
Of Orpheus echo from the Underworld . . .
All Earth's calm landscape shimmers; rainbows dance
Above the mountain meadows wherein Love's
Flocks graze. . . . But what chill shadow, not of cloud,
Is this that darkens noonday's crystal? Whence
Comes that far wail of mourning through the groves?

THE FABULOUS GLASS

For Blanche Reverchon-Jouve

In my deep Mirror's blindest heart
A Cone I planted there to sprout.
Sprang up a Tree tall as a cloud
And each branch bore a loud-voiced load
Of Birds as bright as their own song;
But when a distant death-knell rang
My Tree fell down, and where it lay
A Centipede disgustingly
Swarmed its quick length across the ground!
Thick shadows fell inside my mind;
Until an Alcove rose to view
In which, obscure at first, there now
Appeared a Virgin and her Child;
But it was horrid to behold
How she consumed that Infant's Face
With her voracious Mouth. Her Dress
Was Black, and blotted all out. Then
A phosphorescent Triple Chain
Of Pearls against the darkness hung
Like a Temptation; but ere long
They vanished, leaving in their place
A Peacock, which lit up the glass
By opening his Fan of Eyes:
And thus closed down my Self-regarding Gaze.

CAMERA OBSCURA

When Summer sifts its first dusts through the mesh
Of twig and tendril that the Spring has spun, again
Splashing with verjuice stains the lanes and avenues down which
The annual lovers stroll towards their bliss;
And when along banks and beaches warming waves
Throw up wet limbs like ingots for the wind to wipe
Dry, the sun's fervid kissing to ignite; when high-
Charged and bruise-coloured clouds, like tight
Emotion-swollen bosoms rising, brew
Intoxicating storm-broth for the night:

Desire's beams, breaking through a furtive aperture
Into the *camera obscura* of my dream,
Flash on that secret and uncensored screen
Flagrant fast-changing frescoes filled
With rearing torso-monoliths, strong tender lines
Of thew and tendon carved in bas-relief, gunmetal shine
Like mist from neck to thighs: unflawed anatomies
Of nakedness too dizzying to envisage long:
Marlowe's Leander, Michaelangelic gods, that young
High-diving Mercury I once cut from a sports-page . . .

Their dark or sparkling heads just out of reach
Of my outstretched and empty questing palm, have faces
Hidden or turned away, unclear or with glass eyes
Impersonal and cryptic as a fortune-teller's orb;
And so that other quarry that Desire
Projects alternately inside my sight's closed lids:
The fragile natural heroines with submissive fard-sweet lips
But icebound opal eyes that my male fires must melt
Into admiring mirrors: female cherubim, are all
Like disembodied birds or beauteous busts on plinths of air.

How can the Janus gaze, pinned living to twin poles,
Like a rare moth with one white wing one black,
Fly ever to the act's clear candle-flame?
Rely on memory to back these makeshift shades
With Love's hard-won diplomas of accomplishment? Regret
For lost accomplices of other Summer nights, whose hands
Articulated more than all their voices (restless winds
Around what clandestine hotels: O moonlit hells!), blows back
With long-held burning breath through eyeholes bored
By image-laden rays, into my isolation-cell . . .

Touch cannot undivide the pinioned heart
Foaming with helpless fury that could not be shared
Or lessened by acceptance; nor can speech mean more
Than tired preliminaries to farewell: which leaves when said
A slow deep-rooted sting. Then let these briefly bared
Bright simulacra starving need brings forth
Out of the void between two wounds unwind
Designs of pure lubricity, and people the short peace
Of celibacy with myths' lucid smiling flesh;
And wraithlike vanish, leaving no scar behind.

APOLOGIA

'Poète et non honnête homme.'
PASCAL

1

It's not the Age,
Disease, or accident, but sheer
Perversity (or so one must suppose),
That pins me to the singularly bare
Boards of this trestle-stage
That I have mounted to adopt the pose
Of a demented wrestler, with gorge full
Of phlegm, eyes bleared with salt, and knees
Knocking like ninepins: a most furious fool!

2

Fixed by the nib
Of an inept pen to a bleak page
Before the glassy gaze of a ghost mob,
I stand once more to face the silent rage
Of my unseen Opponent, and begin
The same old struggle for the doubtful prize:
Each stanza is a round, and every line
A blow aimed at the too elusive chin
Of that Oblivion which cannot fail to win.

3

Before I fall
Down silent finally, I want to make
One last attempt at utterance, and tell
How my absurd desire was to compose
A single poem with my mental eyes
Wide open, and without even one lapse
From that most scrupulous Truth which I pursue
When not pursuing Poetry.—Perhaps
Only the poem I can never write is *true*.

THE WRITER'S HAND

What is your want, perpetual invalid
Whose fist is always beating on my breast's
Bone wall, incurable dictator of my house
And breaker of its peace? What is your will,
Obscure uneasy sprite: where must I run,
What must I seize, to win
A brief respite from your repining cries?

Is it a star, the passionate short spark
Produced by friction with another's flesh?
You ache more darkly after. Is it power:
To snap restriction's leash, to leap
Like bloodhounds on the enemy? There is no grip
Can crush the fate you fight. Or is it to escape
Into the dream-perspectives maps and speed create?

You never listen, disillusion's dumb
To your unheeding ear. But see my hand,
The only army to enforce your claim
Upon life's hostile land: five pale, effete,
Aesthetic-looking fingers, whose chief feat
Is to trace lines like these across a page:
What small relief can they bring to your siege!

TO A CONTEMPORARY

You screwed your heart up to incredible
Rigidity; upon your sleeve it glittered like
A jewelled watch tick-tocking. All your wits
Were tough as wire since you, cut to the quick
By premature cold disabuse,
Had set your face against your inmost face
(Which wept, but which no tears could slake).

Inconsolable one, I watched your eyes
(Which never looked in mine), and saw
How often in those mirrors like the stain
Of some white poison slowly spread,
Making all sanguine colour drain

Out of what they reflected of the world outside,
Your ceaseless sense of the ubiquitous Inane.

And when you pinned up on your mouth that smile
Of purest malice by which you betrayed
Your total lack of trust, how all too well
I recognized its likeness to my own twitch of disgust
With mankind and myself . . . (Had I not made
The same unseeing trek through just such cruel
Subjective labyrinths as your lost feet trod?)

Through even your ignominy one saw at last
That finally despairing pride
From which you drew your courage to endure
The worst self-torments of perversity
(The treadmill of your vice,
The automatic all-dismissing sneer,
The quite deliberate invocation of the Void).

Yours was the courage not to turn away
From knowledge or from Death (whose wiles
And ironies by now surely you know
By heart); and to make unbelief
Your only refuge. You were brave
Enough to bear the seeming truth, could you not dare
To face the last fear, which is that of Love?

AN ELEGY

ROGER ROUGHTON 1916–41

Friend, whose unnatural early death
In this year's cold, chaotic Spring
Is like a clumsy wound that will not heal:
What can I say to you, now that your ears
Are stoppered-up with distant soil?
Perhaps to speak at all is false; more true
Simply to sit at times alone and dumb
And with most pure intensity of thought
And concentrated inmost feeling, reach
Towards your shadow on the years' crumbling wall.

I'll say not any word in praise or blame
Of what you ended with the mere turn of a tap;
Nor to explain, deplore nor yet exploit
The latent pathos of your living years—
Hurried, confused and unfulfilled—
That were the shiftless years of both our youths
Spent in the monstrous mountain-shadow of
Catastrophe that chilled you to the bone:
The certain imminence of which always pursued
You from your heritage of fields and sun . . .

I see your face in hostile sunlight, eyes
Wrinkled against its glare, behind the glass
Of a car's windscreen, while you seek to lose
Yourself in swift devouring of white roads
Unwinding across Europe or America;
Taciturn at the wheel, wrapped in a blaze
Of restlessness that no fresh scene can quench;
In cities of brief sojourn that you pass
Through in your quest for respite, heavy drink
Alone enabling you to bear each hotel night.

Sex, Art and Politics: those poor
Expedients! You tried them each in turn,
With the wry inward smile of one resigned
To join in every complicated game
Adults affect to play. Yet girls you found
So prone to sentiment's corruptions; and the joy
Of sensual satisfaction seemed so brief, and left
Only new need. It proved hard to remain
Convinced of the Word's efficacity; or even quite
Certain of World-Salvation through 'the Party Line' . . .

Cased in the careful armour that you wore
Of wit and nonchalance, through which
Few quizzed the concealed countenance of fear,
You waited daily for the sky to fall;
At moments wholly panic-stricken by
A sense of stifling in your brittle shell;
Seeing the world's damnation week by week
Grow more and more inevitable; till
The conflagration broke out with a roar,
And from those flames you fled through whirling smoke,

To end at last in bankrupt exile in
That sordid city, scene of *Ulysses*; and there,
While War sowed all the lands with violent graves,
You finally succumbed to a black, wild
Incomprehensibility of fate that none could share . . .
Yet even in your obscure death I see
The secret candour of that lonely child
Who, lost in the storm-shaken castle-park,
Astride his crippled mastiff's back was borne
Slowly away into the utmost dark.

FROM A DIARY

Imperfections of substance, dross of the day-by-day;
Banality, unlove and disappointment . . . Grey

Webs of attrition, and the trivial tick
Of the nerves' run-down clock—dank skeins of thick

Colourless thought unravelling through the skull,—
This bitter grit of conscience, and the dull

Pulse of internal scars . . . Compression: no
Inscape or scope or space: only the flow

Of stupor's steady muffled fugue.—At night,
While time pursues unwatched its weightless flight,

Blackness lolls on the air, as still as gas
And denser, round each building's lonely mass

Collapsing in the depths of its own dream;
Silence suppresses every pent-up latent scream;

And I lie like a log (as I have lain
How many year-long nights?) and once again,

Immobile, mute, locked in my private room,
Hear, ruminating on the unwritten doom

Awaiting all men's hearts in their dumb solitude,
Within me my heart's numb, indifferent blood.

ODEUR DE PENSÉE

Thought has a subtle odour: which is not
Like that which hawthorn after rainfall has;
Nor is it sickly or astringent as
Are some scents which round human bodies float,
Diluting sweat's thick auras. It's not like
Dust's immemorial smells, which lurk
Where spiders nest, in shadows under doors
Of rooms where centuries have died, and rest
In clouds along the blackening cracked floors
Of sties and closets, attics and wrecked tombs . . .
Thought's odour is so pale that in the air
Nostrils inhale, it disappears like fire
Put out by water. Drifting through the coils
Of the involved and sponge-like brain it frets
The fine-veined walls of secret mental cells,
Brushing their fragile fibre as with light
Nostalgic breezes: And it's then we sense
Remote presentiment of some intensely bright
Impending spiritual dawn, of which the pure
Immense illumination seems about to pour
In upon our existence from beyond
The edge of Knowing! But of that obscure
Deep presaging excitement shall remain
Briefly to linger in dry crannies of the brain
Not the least breath when fear-benumbed and frail
Our dying thought within the closely-sealed
Bone casket of the skull has flickered out,
And we've gone down into the odourless black soil.

FÊTE

After long thirst for sky, there was the sky,
That ether lake: vast azure canopy
Intensely stretched between horizons' ends!
Along the quays
The panes of opening windows flashed like wings,
Weaving long rays among the leafless trees;
Sirens of drifting barges sang:
And the whole day
Drank in the fecund flowing of the sky.

And on the outskirts of the town
Where the last house-blocks take their vacant stare
Across the straggling zone, and rusty streams
Among brown squares of threadbare soil
Persist their irrigating ooze, a savage train
Tore through a cutting with triumphant screams,
Releasing streamers of thick whirling breath
Which climbed and were suspended like presentiments on high . . .

Once more the earth, its buried spirit stirred,
Aspired towards the Summer's splendid bursting
And an illustrious death.

PARIS, 1938

CHAMBRE D'HÔTEL

While a sad Sunday's silver light
Slid through the rain of afternoon
 And slimed the town's grey stone,
We side-by-side without a word
Above the cobbled island quays
Round which rolled on the swollen Seine,
 Lay staring at a white
And barren ceiling: till it seemed
We'd lain forever thus entombed
 Deep in unspeaking spleen.

Oh, when at last I tried to take
Your hand in mine, your stranger's face
 Towards my mouth to bend,
You sprang up from the bed and went
Away, across the room, to stand
And watch, through muslin'd window-glass
 The plane-trees lean to ask
The river what you too asked then,
A riddle without answer and
 As old as earth's disgrace.

JARDIN DU PALAIS ROYAL
To B. Von M.

The sky's a faded blue and taut-stretched flag
Tenting the quadrangle. On three
Sides the arcade (tenebrous lanes
Down which at times patchouli'd ghosts flit by—
Furtive reflections on the filmy panes
Of shops which seem to store only the dusts
And atmospheres of antiquated years,—
Intent on fusty vice), restricts the garden-
Statues' timeless gaze. Here inside this
Shut-off and bygone place, brown urchin birds
Play tag and twitter, jittering around
The central fountain's dance; while children chase
Their ragged shadows round about
The palinged trees, with screams; and iron chairs
With pattern-perforated seats drop their design
Like black lace on the gravel. There we sat
And watched that liquid trembling spire the wind
Made sway and break and spatter a thin spray
Like tears upon our hair and tight-clenched hands . . .
How long? I have forgotten. But you rocked
Backwards and forwards, scraping up small stones,
And never spoke. The day was in July,
Full of a whitish and exhausting glare. And I
Could only stare in silence, trying to see
Into the constantly disintegrating core
Round which the fountain ever climbed again;
Hearing the clack of feet that died away
Down the dim passage, and the unnerving din
Child-voices made behind us. O but then
You turned, and asked me with inconsolable eyes
The meaning of the pain that kept us dumb;
And then we both knew that our pact had been betrayed;
And that cold moment made the garden seem
Too like our lives, abandoned in a wilderness of Time,
Boxed-in by the frustrating and decayed
Walls of the haunted Memory's arcade.

NOCTAMBULES

Hommage à Djuna Barnes

They stand in doorways; then
Step out into the rain
Beneath the lamplight's blue
Aurora; down the street
Towards a blood-red sign
Scrawled swiftly on the wet
Slate of the midnight sky
And then sponged off again . . .
With watchful masks they wait
On stools at bars. I can-
Not see their faces; some
Are weeping; now I hear
A shadow sigh: *The band*
Plays recklessly away
Our last hours, one by one . . .
And then a girl in tulle
With black moths fluttering in
The gold mist of her hair
Enters the hard white pool
Of a great arc-lamp's glare
Revealing, where her face
Should be, a gaping hole!
Their mingling voices roar . . .
Now they have gone again:
The Rue Fontaine is full
Of other shadows; rain
Trickles down postered walls,
Down cafés' plate-glass panes.
Whispers outside the door,—
Words an accordion drowns . . .
Now like the clink of ice
In highball glasses come
Their voices from afar:
Straying from place to place,
Not knowing where we go,
We stumble through our dream
Beneath an evil star . . .
Words the wind's echoes blur,
Lost among tossing trees
Along the Rue Guynemer
Where as the wheezing chimes

Of Ste Sulpice strike three,
In his tight attic high
Above the street, a boy
With a white face which dreams
Have drained of meaning, writes
The last page of a book
Which none will understand:
While down the corridor
Outside the room return
Their faint footsteps again . . .
They wait outside the door;
Their whispers fall like sand
In hour-glasses; I hear
Passionate sobbing; then
A voice that I've heard before
On many a night like this—
Strident with anguish—cries:
Darkness erodes the hearts
Locked in our breasts: the Night
Is gnawing our lives away:
O let Lust deaden without end
This aching void within . . .
And when the voice has died
Away, more cries are heard
Which, merging with the wind
In wordless tumult, blend
In an inconsolable dirge
And desperately press
Onwards in waves across
Acres of wet roofs, on
Across the unseen Seine,
Away beyond the Madeleine
And deep into the gulf that yawns
Behind the Sacré Coeur . . .
The rustling driven rain
Ceases awhile; the air
Hangs numb; Night still wears on.
Now down the desolate wide glade
Of Boulevard Sebastopol,
Beneath the creaking iron boughs
Of shop signs hung along each side,
A young American, intent
On finding a chance bed-fellow,
Pursues a vagrant *matelot's*

Slim likely-looking form . . .
An English drunkard sits alone
In a small *bistro* in Les Halles
And keeps rehearsing the Lord's Prayer
In a mad high-pitched monotone
To the blue empty air.
And in a Left-bank café where
At about half-past four
Exiles are wont to bare
Their souls, a son-and-heir
Of riches and neurosis casts
His frail befuddled blonde
Brutally to the floor
And with despairing fists
Tries to blot out the gaze
Of her wet senseless eyes . . .
One who has wandered long
Through labyrinths of his own brain
More solitary and obscure
Than any maze of stone
Pavements and lamplit walls
Now stops beside the Seine
And leaning down to peer
Into the swirling gloom
Of swollen waters, says:
What day can ever end
The night of those from whom
God turns away his face,
Or what ray's finger pierce
The depths wherein they drown?
Exhaustion brings no peace
To the lost soul . . . But soon
Behind the Eastern slums
A chalky streak of dawn-
Light gradually gleams;
And men from women turn
Away to face the wall,
All lust exhausted, in
Dozens of one-night rooms . . .
Then suddenly a chill
Breath sneaks along the stones
Of narrow streets and makes
The lids of rubbish-bins
To clatter faintly, shakes

The rags and scraps and tins
Strewn in the gutters; and
A rapid shiver runs
Throughout the still, grey, blind
Mass of the city.—Now
As countless times before
I make my roomward way
Across that silent square
Where always as I pass
Them snarling lions stare
At me with stony eyes
From round about the base
Of their dry fountain . . . O!
How derelict is this
Hour of Night's ending: when
The Dark's pale denizens must go
With tales untold and tears
Unwept,—their shrivelled souls
Unsold, unsaved,—back to
The caves of sleep, their worn-
Out beds in lonely holes
Wherein they hide by day.
And climbing the last stair
How timeless seems this time
Of vigil in despair:
Of night by night the same
Weary anabasis
Between two wars, towards
The Future's huge abyss.

SONNET: THE UNCERTAIN BATTLE

Away the horde rode, in a storm of hail
And steel-blue lightning. Hurtled by the wind
Into their eardrums from behind the hill
Came in increasing bursts the startled sound
Of trumpets in the unseen hostile camp.—
Down through a raw black hole in heaven stared
The horror-blanched moon's eye. Across the swamp
Five ravens flapped; and the storm disappeared
Soon afterwards, like them, into that pit
Of Silence which lies waiting to consume

Even the braggart World itself at last . . .
The candle in the hermit's cave burned out
At dawn, as usual.—No one ever came
Back down the hill, to say which side had lost.

LINES

So much to tell: so measurelessly more
Than this poor rusting pen could ever dare
To try to scratch a hint of . . . Words are marks
That flicker through men's minds like quick black dust;
That falling, finally obliterate the faint
Glow their speech emanates. Too soon all sparks
Of vivid meaning are extinguished by
The saturated wadding of Man's tongue . . .
And yet, I lie, I lie:
Can even Omega discount
The startling miracle of human song?

TIME AND PLACE:

'Au temps où la douceur
Est cruelle et le désespoir est brilliant.'
                                    PIERRE JEAN JOUVE

SNOW IN EUROPE

Out of their slumber Europeans spun
Dense dreams: appeasement, miracle, glimpsed flash
Of a new golden era; but could not restrain
The vertical white weight that fell last night
And made their continent a blank.

Hush, says the sameness of the snow,
The Ural and the Jura now rejoin
The furthest Arctic's desolation. All is one;
Sheer monotone: plain, mountain; country, town:
Contours and boundaries no longer show.

The warring flags hang colourless a while;
Now midnight's icy zero feigns a truce
Between the signs and seasons, and fades out
All shots and cries. But when the great thaw comes,
How red shall be the melting snow, how loud the drums!

CHRISTMAS, 1938

ZERO

SEPTEMBER, 1939

Who can by now not hear
The hollow and annihilating roar
Of final disillusion; or not know
How our condition is uncertain and obscure
And difficult to bear? Yet through
The blackness of his dungeon there still peer
Man's eyes, unmoving, lit by their desire
To see *the worst*, and yet not die
Of their lucid despair
But in such vision persevere
Through time into Eternity.
For this is Zero-hour
When the most penetrating gaze can see
Only the Void, the emptier than air,
The incoherent *Nada* of the seer:
Who blind is yet not blind, being aware
Of the Negation's double mystery!

Tomb of what was, womb of what is to be.

AN AUTUMN PARK

Dark suffocates the world; but such
Ubiquity of shadow is unequal. Here
At the spiked gates which crown the hill begins
A reign as of suspense within suspense:
Outside our area of sand-bagged mansions and of tense
But inarticulate expectancy of roars,
The unhistoric park
Extends indifference through all its air.

123

During these present days
None but the lonely and reflective care to walk
Through the unworldly and concealed preserves
Of vegetable integrity (where trees
Though murmurous at least are without words . . .)
For such unsocial ones the park negates
With its consistently non-human peace
All the loud mind-polluted world outside its gates.

When sudden sunrays break the brooding haze
Which makes monotonous these grounds,
Livid the little wind-flaked lakes appear,
Vivid the fever-mottled leaves still bound
By mouldering stalks to idly shaken boughs;
Brief light and breath intensify the scene
With glitter drifting across wet grass wastes
And odour of crushed bracken and raw sand . . .

These acres bordering on plains of brick
And brain and coin and newspaper and noise,
Still store for townsmen such as seek
Remembrance of the simpler earth that was
Our dwelling and contentment once, a chance
Of re-beholding that lost innocence; may show
To those that walk today there to forget, the true
And imminent glory breaking through Man's circumstance.

OCTOBER, 1939

FAREWELL CHORUS

I

And so! the long black pullman is at last departing, now,
After those undermining years of angry waiting and cold tea;
And all your small grey faces and wet hankies slide away
Backwards into the station's cave of cloud. And so Goodbye
To our home-town, so foreign now its lights no longer show;
And to old lives already indistinct as a dull play
We saw while staying somewhere in the Midlands long ago.

Farewell to the few and to the many; for tonight
Our souls may be required of us; and so we say Adieu

To those who charmed us with their ever ready wit
But could not see the point; to those whose polished hands
And voices could allay a little while our private pain
But could not stay to soothe us when worse bouts began;
To those whose beauties were too brief: Farewell, dear friends.

To you as well whom we could never love, hard though
We tried, because our pity told us you were weak,
And because of pity we abhorred; to you
Whose gauche distress and badly-written postcards made us ache
With angrily impatient self-reproach; you who were too
Indelicately tender, whose too soft eyes made us look
(Against our uncourageous wish) swiftly away . . .

To those, too, whom we hardly knew, or could not know;
To the indifferent and the admired; to the once-met
And long-remembered faces: Yes, Goodbye to you
Who made us turn our heads to look again, and wait
For hours in vain at the same place next day;
Who for a moment might have been the lost selves sought
Without avail, and whom we know we never shall find now.

Away, away! Yet now it is no longer in retreat
That we are leaving. All our will is drowned
As by an inner tidal-wave that has washed our regret
And small fears and exhausted implications out of mind.
You can't accompany our journey. Nor may we return
Except in unimpassioned recollections from beyond
That ever-nearer frontier that our fate has drawn.

II

And so let's take a last look round, and say Farewell to all
Events that gave the last decade, which this New Year
Brings to its close, a special pathos. Let us fill
One final fiery glass and quickly drink to 'the Pre-War'
Before we greet 'the Forties', whose unseen sphinx-face
Is staring fixedly upon us from behind its veil;
Drink farewell quickly, ere the Future smash the glass.

Even while underneath the floor are whirling on
The wheels which carry us towards some Time-to-Come,
Let us perform this hasty mental rite (as one
Might cast a few imagined bays into the tomb

125

Of an unloved but memorable great man);
Soon the still-near will seem remotely far; there's hardly time
For much oration more than mere Goodbye, again:

To the delusive peace of those disintegrating years
Through which burst uncontrollably into our view
Successive and increasingly premonitory flares,
Explosions of the dangerous truth beneath, which no
Steel-plated self-deception could for long withstand . . .
Years through the rising storm of which somehow we grew,
Struggling to keep an anchored heart and open mind,

Too often failing. Years through which none the less
The coaxing of complacency and sleep could still persuade
Kind-hearted Christians of the permanence of Peace,
Increase of common-sense and civic virtue. Years which bade
Less placid conscientious souls indignantly arise
Upon ten thousand platforms to proclaim the system mad
And urge the liquidation of a senile ruling-class.

Years like a prison-wall, frustrating though unsound
On which the brush of History, with quick, neurotic strokes,
Its latest and most awe-inspiring fresco soon outlined:
Spenglerian lowering of the Western skies, red lakes
Of civil bloodshed, free flags flagrantly torn down
By order of macabre puppet orators, the blind
Leading blindfolded followers into the Devil's den . . .

III

And so, Goodbye, grim 'Thirties. These your closing days
Have shown a new light, motionless and far
And clear as ice, to our sore riddled eyes;
And we see certain truths now, which the fear
Aroused by earlier circumstances could but compromise,
Concerning all men's lives. Beyond despair
May we take wise leave of you, knowing disasters' cause.

Having left all false hopes behind, may we move on
At a vertiginous unmeasured speed, beyond, beyond,
Across this unknown Present's bleak and rocky plain;
Through sudden tunnels; in our ears the wind
Echoing unintelligible guns. Mirrored within
Each lonely consciousness, War's world seems without end.
Dumbly we stare up at strange skies with each day's dawn.

Could you but hear our final farewell call, how strained
And hollow it would sound! We are already far
Away, forever leaving further leagues behind
Of this most perilous and incoherent land
We're in. The unseen enemy are near.
Above the cowering capital Death's wings impend.
Rapidly under ink-black seas today's doomed disappear.

We are alone with one another, but our eyes
Meet seldom in the dark. What a relentless roar
Stuffs every ear, as though with wool! The winds that rise
Out of our dereliction's vortex, hour by hour,
To bring us word of the incessant wordless guns,
Tirades of the insane, thick hum of planes, the rage of fire,
Eruptions, waves: all end in utmost silence in our brains.

'The silence after the viaticum.' So silent is the ray
Of naked radiance that lights our actual scene,
Leading the gaze into those nameless and unknown
Extremes of our existence where fear's armour falls away
And lamentation and defeat and pain
Are all transfigured by acceptance; where men see
The tragic splendour of their final destiny.

NEW YEAR, 1940

SPRING MCMXL

London Bridge is falling down, Rome's burnt, and Babylon
The Great is now but dust; and still Spring must
Swing back through Time's continual arc to earth.
Though every land become as a black field
Dunged with the dead, drenched by the dying's blood,
Still must a punctual goddess waken and ascend
The rocky stairs, up into earth's chilled air,
And pass upon her mission through those carrion ranks,
Picking her way among a maze of broken brick
To quicken with her footsteps the short sooty grass between;
While now once more their futile matchwood empires flare and
    blaze
And through the smoke men gaze with bloodshot eyes
At the translucent apparition, clad in trembling nascent green,
Of one they can still recognize, though scarcely understand.

A WARTIME DAWN

Dulled by the slow glare of the yellow bulb;
As far from sleep still as at any hour
Since distant midnight; with a hollow skull
In which white vapours seem to reel
Among limp muddles of old thought; till eyes
Collapse into themselves like clams in mud . . .
Hand paws the wall to reach the chilly switch;
Then nerve-shot darkness gradually shakes
Throughout the room. *Lie still* . . . Limbs twitch;
Relapse to immobility's faint ache. And time
A while relaxes; space turns wholly black.

But deep in the velvet crater of the ear
A chip of sound abruptly irritates.
A second, a third chirp; and then another far
Emphatic trill and chirrup shrills in answer; notes
From all directions round pluck at the strings
Of hearing with frail finely-sharpened claws.
And in an instant, every wakened bird
Across surrounding miles of air
Outside, is sowing like a scintillating sand
Its throat's incessantly replenished store
Of tuneless singsong, timeless, aimless, blind.

Draw now with prickling hand the curtains back;
Unpin the blackout-cloth; let in
Grim crack-of-dawn's first glimmer through the glass.
All's yet half sunk in Yesterday's stale death,
Obscurely still beneath a moist-tinged blank
Sky like the inside of a deaf mute's mouth . . .
Nearest within the window's sight, ash-pale
Against a cinder coloured wall, the white
Pear-blossom hovers like a stare; rain-wet
The further housetops weakly shine; and there,
Beyond, hangs flaccidly a lone barrage-balloon.

An incommunicable desolation weighs
Like depths of stagnant water on this break of day.—
Long meditation without thought.—Until a breeze
From some pure Nowhere straying, stirs
A pang of poignant odour from the earth, an unheard sigh

Pregnant with sap's sweet tang and raw soil's fine
Aroma, smell of stone, and acrid breath
Of gravel puddles. While the brooding green
Of nearby gardens' grass and trees, and quiet flat
Blue leaves, the distant lilac mirages, are made
Clear by increasing daylight, and intensified.

Now head sinks into pillows in retreat
Before this morning's hovering advance;
(Behind loose lids, in sleep's warm porch, half hears
White hollow clink of bottles,—dragging crunch
Of milk-cart wheels,—and presently a snatch
Of windy whistling as the newsboy's bike winds near,
Distributing to neighbour's peaceful steps
Reports of last-night's battles); at last sleeps.
While early guns on Norway's bitter coast
Where faceless troops are landing, renew fire:
And one more day of War starts everywhere.

APRIL, 1940

WALKING AT WHITSUN

'La fontaine n'a pas tari
Pas plus que l'or de la paille ne s'est terni
Regardons l'abeille
Et ne songeons pas à l'avenir...'
APOLLINAIRE

... Then let the cloth across my back grow warm
Beneath such comforting strong rays! new leaf
Flow everywhere, translucently profuse,
And flagrant weed be tall, the banks of lanes
Sprawl dazed with swarming lion-petalled suns
As with largesse of pollen-coloured wealth
The meadows; and across these vibrant lands
Of Summer-afternoon through which I stroll
Let rapidly gold glazes slide and chase
Away such shades as chill the hillside trees
And make remindful mind turn cold...

                    The eyes
Of thought stare elsewhere, as though skewer-fixed
To an imagined sky's immense collapse;
Nor can, borne undistracted through this scene
Of festive plant and basking pastorale,
The mind find any calm or light within
The bone walls of the skull; for at its ear
Resound recurrent thunderings of dark
Smoke-towered waves rearing sheer tons to strike
Down through Today's last dyke. Day-long
That far thick roar of fear thuds, on-and-on,
Beneath the floor of sense, and makes
All carefree quodlibet of leaves and larks
And fragile tympani of insects sound
Like Chinese music, mindlessly remote,
Drawing across both sight and thought like gauze
Its unreality's taut haze.

                    But light!
O cleanse with widespread flood of rays the brain's
Oppressively still sickroom, wherein brood
Hot festering obsessions, and absolve
My introspection's mirror of such stains
As blot its true reflection of the world!
Let streams of sweetest air dissolve the blight
And poison of the News, which every hour
Contaminates the ether.

                    I will pass
On far beyond the village, out of sight
Of human habitation for a while.
Grass has an everlasting pristine smell.
On high, sublime in his bronze ark, the sun
Goes cruising across seas of silken sky.
In fields atop the hillside, chestnut trees
Display the splendour of their branches piled
With blazing candle burdens.—Such a May
As this might never come again . . .

                    I tread
The white dust of a weed-bright lane; alone
Upon Time-Present's tranquil outmost rim,
Seeing the sunlight through a lens of dread,
While anguish makes the English landscape seem

Inhuman as the jungle, and unreal
Its peace. And meditating as I pace
The afternoon away, upon the smile
(Like that worn by the dead) which Nature wears
In ignorance of our unnatural tears,
From time to time I think: How such a sun
Must glitter on their helmets! How bright-red
Against this sky's clear screen must ruins burn ...

How sharply their invading steel must shine!

MARSHFIELD, MAY, 1940

OXFORD: A SPRING DAY

For Bill

The air shines with a mild magnificence ...
Leaves, voices, glitterings ... And there is also water
Winding in easy ways among much green expanse,

Or lying flat, in small floods, on the grass;
Water which in its widespread crystal holds the whole soft song
Of this swift tremulous instant of rebirth and peace.

Tremulous—yet beneath, how deep its root!
Timelessness of an afternoon! Air's gems, the walls' bland grey,
Slim spires, hope-coloured fields: these belong to no date.

1941

THE GRAVEL-PIT FIELD

Beside the stolid opaque flow
Of rain-gorged Thames; beneath a thin
Layer of early evening light
Which seems to drift, a ragged veil,
Upon the chilly March air's tide:
Upwards in shallow shapeless tiers
A stretch of scurfy pock-marked waste
Sprawls laggardly its acres till
They touch a raw brick-villa'd rim.

131

Amidst this nondescript terrain
Haphazardly the gravel-pits'
Rough-hewn rust-coloured hollows yawn,
Their steep declivities away
From the field-surface dropping down
Towards the depths below where rain-
Water in turbid pools stagnates
Like scraps of sky decaying in
The sockets of a dead man's stare.

The shabby coat of coarse grass spread
Unevenly across the ruts
And humps of lumpy soil; the bits
Of stick and threads of straw; loose clumps
Of weeds with withered stalks and black
Tatters of leaf and scorched pods: all
These intertwined minutiae
Of Nature's humblest growths persist
In their endurance here like rock.

As with untold intensity
On the far edge of Being, where
Life's last faint forms begin to lose
Name and identity and fade
Away into the Void, endures
The final thin triumphant flame
Of all that's most despoiled and bare:
So these least stones, in the extreme
Of their abasement might appear

Like rare stones such as could have formed
A necklet worn by the dead queen
Of a great Pharaoh, in her tomb . . .
So each abandoned snail-shell strewn
Among these blotched dock-leaves might seem
In the pure ray shed by the loss
Of all man-measured value, like
Some priceless pearl-enamelled toy
Cushioned on green silk under glass.

And who in solitude like this
Can say the unclean mongrel's bones
Which stick out, splintered, through the loose
Side of a gravel-pit, are not

The precious relics of some saint,
Perhaps miraculous? Or that
The lettering on this Woodbine-
Packet's remains ought not to read:
Mene mene tekel upharsin?

Now a breeze gently breathes across
The wilderness's cryptic face;
The meagre grasses scarcely stir;
But when some stranger gust sweeps past,
Seeming as though an unseen swarm
Of sea-birds had disturbed the air
With their strong wings' wide stroke, a gleam
Of freshness hovers everywhere
About the field: and tall weeds shake,

Leaves wave their tiny flags to show
That the wind blown about the brow
Of this poor plot is nothing less
Than the great constant draught the speed
Of Earth's gyrations makes in Space . . .
As I stand musing, overhead
The zenith's stark light thrusts a ray
Down through dusk's rolling vapours, casts
A last lucidity of day

Across the scene: and in a flash
Of insight I behold the field's
Apotheosis: No-man's-land
Between this world and the beyond,
Remote from men and yet more real
Than any human dwelling-place:
A tabernacle where one stands
As though within the empty space
Round which revolves the Sage's Wheel.

SPRING, 1941

REQUIEM

'Permets que nous te goûtions d'abord le jour de la mort
Qui est un grand jour de calme d'épousés,
Le monde heureux, les fils réconciliés.'

PIERRE JEAN JOUVE

I

[Voice: recitative]

O hidden Face! O gaze fixed on us from afar
And that we cannot meet: Grant us, who wait
In the great park of crumbling monuments that is
The world, that we may meet at last those eyes
In which black fires burn back to white,
With perfect clearness, and not blurred by fever's heat
Nor in the sudden spasm of disintegrating fear
That rends the breast of beasts and blinds
The blind and undefined: And O instruct
Us how to ripen unto Thee.

[Choir: sotto voce]

Hearts are unripe
And spirits light as straw that in Thy light
Shall kindle like the straw, and flare away
To nothing in an instant breath of smoke.

[Voice]

Thy light is like a darkness and Thy
Joy is found through grief. And they who search
For Thee shall find Thee not. And hidden in Thy mouth
The blinding benediction of the final phrase
Which shall not fall upon a listening ear.

[Choir]

For they who listen at the secret door
Hear only their own heart beat out its fault.

II

[Voice]

In the great park,
A wanderer at sundown by the weeping falls
Of pallid spume and high prismatic spray
Once saw across the water in the last illusive light
A figure with a gleaming chalice come . . .

134

[*Choir*]
But it was not Thy Angel!

[*Voice*]
                    And another heard
A warning echo in a mountain cave,
Reverberant with distance and the undertone of guilt . . .

[*Choir*]
But it was not Thy voice!

[*Voice*]
                    For silent and invisible
Are all Thy works; and hidden in the depths midway between
Desire and fear. And they who long for Thee and are afraid
Of Life, and they who fear the clear stroke of Thy knife
Obsessed with the pale shadows of themselves, shall lose full sight
And understanding of that final mystery.

III

[*Choir*]
   Tenebral treasure and immortal flower
   And flower of immortal Death!
   O silent white extent
   Of skyless sky, the wingless flight
   And the long flawless cry
   Of aspiration endlessly!

[*Voice*]
The seed is buried in us like a memory; the seed
Is hidden from us like the omnipresent Eye; it grows
Within us through Time's flux, both night and day.

[*Choir*]
Darkness that burns like light, black light
And essence of all radiance!
O depth beyond confusion sunk,
The timeless Nadir at the heart
Of Time, where all creative and
Destructive forces meet!

The seed is nurtured by involuntary tears; by blood
Shed from Love's inmost wounds; its roots are fed
By the concealed corruption of unknown desires.

[Choir]

   We cannot hear or see, nor say
   The name: There is no light
   Or shade, nor place nor time,
   No movement, no repose,
   But only perfect prescience
   Of the Becoming of the Whole.

[Voice]

The seed springs from us into flower; yet none can tell
At what hour late or early those concealed furled leaves
And multifoliate petals shall outgrow their tender shell.

[Choir]

   The hour is unknown:
   The hour endures:
   The hour strikes every hour.

IV

[Voice]

Each hour of life is glorious and vain.
O thirst and glorious unsatisfied
Lamenting cry! How vain the short relief
And unabiding refuge from the tide
That nearer crawls each day across the sands
On which our house is founded! Vanity
Of vanities, all things held by our hands!
Beyond their reach, with diamond-rays, and high
Above the furthest fields of ether lies
The core of glory, only ascertained
By inward opening of Death's deep eye
And outward flight of Spirit long sustained:

   [Choir: distantly echoing]

By wings the swift flames of the funeral pile
Are fanned . . . Dead faces guard a secret smile.

1938–40

ELSEWHERE

La vraie vie est ailleurs . . .
RIMBAUD

Profound is inexistence on this earth
   Among our human kind:
                Profound
The weight of absence on the sleeping heart
That all war's detonations cannot rouse:
Rumour of selfless hordes with eyes
Red-rimmed and haggard, swarming through the dirt
Of ruined palaces: the roar
Of cannon-mouths, of sawtoothed mouths, the mouths
Of printing-presses, megaphonic maws
Of the possessed and the psychotic: and the pounding waves
Of automatic labour on the daily shore:
Rocked by this deep
And oil-black ocean's tidal pulse
The stunned soul sleeps,
Profoundly absent from its body's condemned house.

The taste of pleasure's now like sand between the teeth;
Worn-out, the nerves are numb; and Death's
Most sumptuous music strikes the ear like wind
Forced dumbly out of emptiness.
                   The sun
Strikes cold upon our nakedness, and shines
With rays of shadow through the diffuse light
Of interstellar space;
While over the last phase of night
The dead face of the moon hangs like a curse.
Deep in our empty sky hangs like a moon
The curse of inexistence; while the spirit sleeps
Profoundly absent from the earth.
                 But on
Negation's further shore, the yonder side
Of sleep and absence, dazzling is the sheer
Rockface set like an ice-barred gate
Beneath that nether tableland's pure height:
Whose sky is the negation of our sky,
Where all earth's ruins are rebuilt
Of stone that sings, and cold fire burns
The scentless incense of the air:
Where time and number are once more atoned
And to its true existence the Unnamed returns.

137

CONCERT OF ANGELS

To Kay Boyle

I

Wind! Out of the night of desolate
negation that we suffer in the waste
of time and impotence of thought,
rise in the mind and out of stupor stir
the hidden hearing with deep
echoes from the spirit host of
angels! Their intensely rapt,
almost inhuman faces luminous
with utmost concentration, the incisive bows
held in their long keen hands—enchanted swords
to slay the earth-binding ear and so release
the lost celestial sense—carving broad curves
across the nerve-taut strings, and like invisible
irradiations of sheer light, like resonance
of huge cathedral bell-notes hovering
over the earth in rings of fiery mist,
their clear cathartic music welling out
into infinity's unfathomed well.

II

While from the sonorous black well
emerge and palely fade and form again
white disembodied hands like drifting flames,
buds, tender leaves and tendrils shaped like hands,
and vision-clouded faces cloudily
impending on the air, with hidden eyes
and hungry mouths like mouths distraught with prayer.

Darkness's mouth, which opens in us now
in the most secret place, is over-brimmed
with straining hymns, with stars
like fountains burning upwards with the impetus
of flying gothic buttresses whose rainbow-arc
both aspiration and sustaining force contains.

III

Here is the transcendental source
of every human cry, replenished by
the deepest chords of death, by shrill

destruction's laughter, by the thrilling
arias of love, so lofty none can tell
or human or divine; and shock-torn sobs
of rape and copulation, exiled sighs,
corrupted beauty's ravishing lament, the long
nostalgic call that answers skylines, transference
from mortal sound into eternal song!

Let there be praise, praise and
praise, organic orchestra and cloudy choir,
to the great incandescent power
of sublimation, vitalizing clay,
with sacred fire consuming the grey dross
of sleep and sickness, balancing
in perfect tension between dark and light
the horrid depth, the spiritual height.

VII
A VAGRANT & OTHER POEMS
(1946–1950)

A VAGRANT

'Mais il n'a point parlé, mais cette année encore
Heure par heure en vain lentement tombera.'
                                    ALFRED DE VIGNY

'They're much the same in most ways, these great cities. Of them
    all,
Speaking of those I've seen, this one's still far the best
Big densely built-up area for a man to wander in
Should he have ceased to find shelter, relief,
Or dream in sanatorium bed; should nothing as yet call
Decisively to him to put an end to brain's
Proliferations round the possibilities that eat
Up adolescence, even years up to the late
Thirtieth birthday; should no one seem to wait
His coming, to pop out at last and bark
Briskly: "A most convenient solution has at last
Been found, after the unavoidable delay due to this spate of wars
That we've been having lately. This is it:
Just fill in (in block letters) on the dotted-line your name
And number. From now on until you die all is
O.K., meaning the clockwork's been adjusted to accommodate
You nicely; all you need's to eat and sleep,
To sleep and eat and eat and laugh and sleep,
And sleep and laugh and wake up every day
Fresh as a raffia daisy!" I already wake each day
Without a bump or too much morning sickness to routine
Which although without order wears the will out just as well
As this job-barker's programme would. His line may in the end
Provide me with a noose with which to hang myself, should I
Discover that the strain of doing nothing is too great
A price to pay for spiritual integrity. The soul
Is said by some to be a bourgeois luxury, which shows
A strange misunderstanding both of soul and bourgeoisie.
The Sermon on the Mount is just as often misconstrued
By Marxists as by wealthy congregations, it would seem.
The "Modern Man in Search of Soul" appears
A comic criminal or an unbalanced bore to those
Whose fear of doing something foolish fools them. *Je m'en fous!*
Blessèd are they, it might be said, who are not of this race
Of settled average citizens secure in their *état*
Civil of snowy guiltlessness and showy high ideals
Permitting them achieve an inexpensive lifelong peace
Of mind, through dogged persistence, frequent aspirin, and bile

143

Occasionally vented via trivial slander . . . Baa,
Baa, O sleepysickness-rotted sheep, in your nice fold
Are none but marketable fleeces. I my lot
Prefer to cast at once away right in
Among the stone-winning lone wolves whose future cells
Shall make home-founding unworthwhile. Unblessèd let me go
And join the honest tribe of patient prisoners and ex-
Convicts, and all such victims of the guilt
Society dare not admit its own. I would not strike
The pose of one however who might in a chic ballet
Perform an apache role in rags of cleverly-cut silk.
Awkward enough, awake, yet although anxious still just sane,
I stand still in my quasi-dereliction, or but stray
Slowly along the quais towards the ends of afternoons
That lead to evenings empty of engagements, or at night
Lying resigned in cosy-corner crow's-nest, listen long
To sounds of the surrounding city desultorily
Seeking in loud distraction some relief from what its nerves
Are gnawed by: I mean knowledge of its lack of *raison d'être*.
The city's lack and mine are much the same. What, oh what can
A vagrant hope to find to take the place of what was once
Our expectation of the Human City in which each man might
Morning and evening, every day, lead his own life, and Mans?'

THE SACRED HEARTH
To George Barker

You must have been still sleeping, your wife there
Asleep beside you. All the old oak breathed: while slow,
How slow the intimate Spring night swelled through those depths
Of soundlessness and dew-chill shadow on towards the day.
Yet I, alone awake close by, was summoned suddenly
By distant voice more indistinct though more distinctly clear,
While all inaudible, than any dream's, calling on me to rise
And stumble barefoot down the stairs to seek the air
Outdoors, so sweet and somnolent, not cold, and at that hour
Suspending in its glass undrifting milk-strata of mist,
Stilled by the placid beaming of the adolescent moon.
There, blackly outlined in their moss-green light, they stood,
The trees of the small crabbed and weed-grown orchard,
Perfect as part of one of Calvert's idylls. It was then,
Wondering what calm magnet had thus drawn me from my bed,

I wandered out across the briar-bound garden, spellbound. Most
Mysterious and unrecapturable moment, when I stood
There staring back at the dark white nocturnal house,
And saw gleam through the lattices a light more pure than gold
Made sanguine with crushed roses, from the firelight that all night
Stayed flickering about the sacred hearth. As long as dawn
Hung fire behind the branch-hid sky, the strong
Magic of rustic slumber held unbroken; yet a song
Sprang wordless from inertia in my heart, to see how near
A neighbour strangeness ever stands to home. George, in the wood
Of wandering among wood-hiding trees, where poets' art
Is how to whistle in the dark, where pockets all have holes,
All roofs for refugees have rents, we ought to know
That there can be for us no place quite alien and unknown,
No situation wholly hostile, if somewhere there burn
The faithful fire of vision still awaiting our return.

INNOCENCE AND EXPERIENCE

Beneath the well-born weak-lined gentle flesh
Its firmly-moulded bonework did much to sustain
This face's actively upheld nobility. I had the time
To gaze upon a late transmuted beauty
Known none too kindly to the North in our cold time.
Yet I knew warmth was there, where were born both
Her Southern mildness and Repression's bleakest whim,
Which is to spoil the good with greatness, till it do its best
To die in surfeit of a passion lean as sin.
I still knew of her nothing less than this,
She could well have played Portia in Spanish
Making it seem a Terry had conceived
To play the cello to a foreign bard's guitar.
Attentive, I beheld a less premeditated look
Melting the mask till one could see it once had worn
The serene, robust air as of never-rebuked gaiety
That shakes like laughter round a regally-loved child;
And saw her clamber up, her will supported
By the arms of his gold braid-adorned dark dignity,
Till safe in peril perching, from the lofty balustrade
She overlooked a square where waved and roared
In passionate approval of political Papa
The population, it appeared, of the then nascent State.

She'd come down to the mezzanine in person
To welcome us, dismissed the footman, stepped
With lifted dress-train held bunched at the knees
Into the ivory-panelled gilt-grilled lift;
Dismissed her maid on reaching the third floor
And shown us down a quite dark passage, hung
With glass-masked pastels—Redon, Morisot, maybe,—
To her most private salon. One could tell
At once how long she must have sat alone,
Sad lady, with the back of her fauteuil
Turned to the uncommunicative view
Of drear palatial faubourg roofs displayed
Between portentous casement draperies,
There in that room the hotel's master had
But seldom entered, though his youth's collections here
As elsewhere were the source of all that caught
The roving eye: a Degas statuette,
A hand-high Rodin piece; upon the wall
Above the fireplace, a nice Géricault—
Two Turkish ladies, or *baigneuses*; some fine
Old pots, and a miraculously carved
Ivory ball within a ball within a ball
That stood upon the escritoire, still piled
With business correspondence that no secretary
Could have availed much to diminish. 'How
Long it must be now since we last—
When was it? Oh, the Occupation? Yes,
I remained here all the time, I held
The fort. A long grim winter. But Eugène,
Of course, had other things to occupy
In South America his busy mind, than my
Predicament. Nothing changed him; simply we
Became "loyally indifferent"; or I trust I so appeared.'

Under the weight of false presuppositions hanging round
Upon all three of us, the other lady frowned (touched too; too
    tired)—
Her constant lit cheroot let fall a not entirely
Inappropriate tiny elegy of ash. Three enigmatic masks.

Outside upon the Plaza, the huge crowd still waved and waved!

'God gives us all, yet no one asks
What it is given for . . .'

PHOTOGRAPH

To Philippe Soupault

Whatever you were looking at when Abbott's camera clicked,
It hardly wore the likeness, I suppose, that you wear now;
Yet its reality can hardly have been other than the one
That we both recognize at present, which is made real
By us and all who truly live in it. Your eyes
Are clear, more clear and keen than what they see, and gaze
  through pain,
Frustration and the future of futility. They look
Straight into the hid heart of whatsoever lies ahead, with active
  trust,
With scepticism and with the tried affection that cannot ever be
Made disappointed by its object's failures. You will thus always be
  aware
That what is true is lovable, and you in knowing this
Will have become one in whose love the love of others may find
  rest.

REPORTED MISSING

At the end of the sunny, polished corridor
I opened a door I had not seen before
And stepped into a room in which the air
Had long been undisturbed but was not stale but
Sleepy sweet and half-familiar, half
Reminiscent of another time and life. There were
Bookshelves and two deep basket chairs, that faced
Each other, though the bed was single, spread
With a soft paisley-patterned cloth, no more to be
Unmade. The view from the dormer window, creeper-fringed,
Was the best in the house. Upon the mantelshelf
Stood lonely in its leather frame a photograph I'll not
Forget, I think, although I never met
The sitter, so immediate was the subjugating charm
That struck one from the eyes and features. These
Reported how much he was missing, whom I cannot praise,
Only commemorate in a few unasked-for lines
Which must leave the essential once more all but quite unsaid.

147

A TOUGH GENERATION

To grow unguided at a time when none
Are sure where they should plant their sprig of trust;
When sunshine has no special mission to endow
With gold the rustic rose, which will run wild
And ramble from the garden to the wood
To train itself to climb the trunks of trees
If the old seedsman die and suburbs care
For sentimental cottage-flowers no more;
To grow up in a wood of rotted trees
In which it is not known which tree will be
First to disturb the silent sultry grove
With crack of doom, dead crackling and dread roar—
Will be infallibly to learn that first
One always owes a duty to oneself;
This much at least is certain: one must live.
And one may reach, without having to search
For much more lore than this, a shrewd maturity,
Equipped with adult aptitude to ape
All customary cant and current camouflage;
Nor be a whit too squeamish where the soul's concerned,
But hold out for the best black market price for it
Should need remind one that one has to live.
Yet just as sweetly, where no markets are,
An unkempt rose may for a season still
Trust its own beauty and disclose its heart
Even to the woodland shade, and as in sacrifice
Renounce its ragged petals one by one.

THE OTHER LARRY

Inwardly corrosive, but to eyes outside most bland,
Chubby and blonde and chuckling: O sardonic friend,
Easily reconciled with, you are sorry after
The black flicked barb has stung
Some tiresome feeble person's too exposed,
Too tender epidermis, though not very and not long:
Exacerbated not yet middle-aged patrician,
Exiled by futile circumstances, ever too well-bred
To make a show of bitterness except in smooth-tongued verse.

Such comment can but seem inept, coming from one
Who's never seen the South of which you sing
But still believes that you will not succeed
In finally convincing all of those
Whom your performance entertains
And makes uncomfortable
That you were meant to grow into a gargoyle
Uttering artful chains of occult smoke-rings
Outside a disbelieved-in anti-god's abode.

EROS ABSCONDITUS

'Wo aber sind die Freunde? Bellarmin
Mit dem Gefahrten . . .'
                                    HÖLDERLIN

Not in my lifetime, the love I envisage:
Not in this century, it may be. Nevertheless inevitable.
Having experienced a foretaste of its burning
And of its consolation, although locked in my aloneness
Still, although I know it cannot come to be
Except in reciprocity, I know
That true love is gratuitous, and will race through
The veins of the reborn world's generations, free
And sweet, like a new kind of electricity.

The love of heroes and of men like gods
Has been for long a strange thing on the earth
And monstrous to the mediocre. They
In whom such love is luminous can but transcend
The squalid inhibitions of those only half alive.
In blind content they breed who never loved a friend.

THE GOOSE-GIRL

She at whose feet I'll finally fall down
With all my niggardly belated offering
Of real emotion, is a lonely silent girl
Who knows no more than I about love's boon
But sits and wonders—feeling at a loss

Among the queens and conquerors who stroll
So poised and pleased about the social scene—
Waiting for no one from an old wives' tale,
But for a childless father and her father's unborn son.

BEWARE BEELZEBUB

Listen, lover of the glistening peril,
The lure lascive and wistful, the sweet pain
Young lacing limbs delight in: the Devil
Will never after smile at you again
When once your easy acquiescence
To his swift-reckoned bargain has put you
Within the power of his swarming lieutenants,
Who lurk in dull disguise the world's mart through
Like fellow fallen men, until the sign
By which the lustless single out a sinner
Bids them to batten, faithful flock of flies,
Dutiful doggers, buzz and drone and whine,
Upon fresh ill-famed flesh for their King's dinner,
Rich-riddled with the worm that never dies.

RONDEL FOR THE FOURTH DECADE

The mind if not the heart turns cold
Seeing the calendar's leaves flying;
Still dare not yet cease trying
To reconcile the heart with growing old.

However often heart's fortune be told
By sceptic mind, the pulse beats on relying
On sanguine heat for hope to hold
Fast to for help when age comes sighing.

But autumn's leaves must cease defying
Grave law and fall like Danae's gold
To stuff blind mouths when, as they turn to mould,
The heart's remains lie still denying
Mind ever knew the truth while dying.

SEPTEMBER SUN: 1947

Magnificent strong sun! in these last days
So prodigally generous of pristine light
That's wasted only by men's sight who will not see
And by self-darkened spirits from whose night
Can rise no longer orison or praise:

Let us consume in fire unfed like yours
And may the quickened gold within me come
To mintage in due season, and not be
Transmuted to no better end than dumb
And self-sufficient usury. These days and years

May bring the sudden call to harvesting,
When if the fields Man labours only yield
Glitter and husks, then with an angrier sun may He
Who first with His gold seed the sightless field
Of Chaos planted, all our trash to cinders bring.

THE POST-WAR NIGHT

No, nowadays at night the flush of light
Reflected anxiously by urban skies, impresses eyes
In quest of soothing space between the stars, as with a sense
Of guilt, not reassurance. This is Peace,
Our nightly black-out dream; yet back to black skies fly
Our eyes disheartened by futility, to seek
Some sterner strength in the unmoonlit midnight's zenith
Above our heads rebuking light's illusions . . . *In our time
We have had vision.* Now our seeing tries
Not to find blindness everywhere it peers,
Relinquishing belief in any sight surpassing this.
*We must see how to justify ourselves
Always.* Perhaps indeed that is for ever all
Our eyes are used to look for: We must stand
Justified:—if not before the whole world then before
Ourselves: if not before the candid inmost heart,
Blandly at least before shrewd common-sense
Sole supreme tribunal in this business-driven world,
Still so remote from all the innate sense

Of human destiny that we are born with knows
To be truly our aim on earth: one God-ruled globe,
Finally unified, at peace, free to create! *That sense
Is dull in all but few today* . . . They are not listened to.
They seldom speak. And how absurd they sound
To such as do hear them, how like a child's
Sublime simplicity and sweet ineptitude,
To talk of Brotherhood and of the beautiful
Smooth-running Great Society that might tomorrow mean
Our paradise regained! How well our guilt,
Long versed in all the necessary lies
Required to run the world in practice knows
How always to remain the same calm, sane
Comfortably compromised collusionists, still safe and sound
At least as long as this false peacetime lasts.

DEMOS IN OXFORD STREET

The Ages of the World, since Adam delved
And Eve remained the perfect lady, still
As innocent of culture as her spouse of apron-string,
Having devolved, have brought us the mature
And really average population passing by, away
And onward down this thoroughfare, of all surely the most
Average in any average modern capital. O Sting!
Where is our life? Where is my neighbour, Love?
We have hardened our faces against each other's weariness
Who walk this way; we are not bound to one another
By bomb panic or famine and it is not Christmas Day.
We are aware of Socialists in power at Westminster
Who seem to be making a pretty mess of things: This evening's *Star*
Has bills that tell of Scandal and Enquiry being made
Much in the interest of the Public (i.e. We,
The People) by such as have its interest at heart . . .
We too, while quite disinterested, have of course got hearts.
The latter are as good as most; but who would dare
Risk giving good away each day with maybe no returns?
Besides, we have our families to think of,
And our families have not got too much to spare
Of time or money, tears or trouble. Stare
As boldly as you like into our faces, we'll not turn
Aside out of your way. We're not the Working-Class.

EVENING AGAIN

Evening again.
             The lurid fuming light
That red sky's smouldering alkali spreads on reflecting stone
Façades of ageing buildings seeming now to slant and strain
Backwards against the leaden East, sheer haggard cliffs
Pitted with windows, baffles with its glare
Those gazing panes. They see nothing but the wrath
Of still prolonged and future conflagrations. With the stain
Of night arising stealthily behind them, fresh leaves shake
Back on their rigid branches, shudder brusquely back and show
How underneath their sparkling green profusion there are hung
Shadows, dull undertone of mourning. Die down, die
Away, brisk wind, let the lit leaves lie still.
Let them with tranquil glitter once more hide
Their secret. Heavy beneath all that is seen
Hangs the forgotten.
                Heavily night falls.
                        When shall I desire
No more for rest from restlessness as evening ends?
When no more into silence sinks the sigh that asks for joy.

THREE VENETIAN NOCTURNES

I. BARCAROLLE

Each blue sun-floodlit day floats through a green evening till Night
Releases flows of indigo to stain sea, sky and shore;
And deep into dark velvet folds are absorbed from the air
The orchestrated murmurs of the crowd and bursts of bright
Abruptly ebbing brassy music bruited from the Square.

On the Lagoon drift shreds of serenade from lanterned boats
That bob more quickly like a pulse when from the Lido steers
Close past them the returning *vaporetto*; the heart beats
More quickly for a moment, lifted on a wave of tears
Upwelling but not breaking in the eyes of one who floats

Reclining in a gondola alone and with the tide
Being borne across the Bacino towards where all the stars

In heaven like spilt pearls blur on the black robe Venice wears
Slackly undulating round her when as a nocturnal bride
She mourns her morning glory long drowned in the sea of years.

2. LIDO GALA FIREWORKS

Rockets released tonight rush up to rape the grapebloom sky:
Rainbows of gelid jewellery smashed to flashlit smithereens
And moulting molten-crystal plumes of birds of paradise
Spontaneously splintering their mixed Murano tints
Into a slowly dropping drift of dust of opals, Milky Way
Stained with a long dynasty of fire-peacocks' last blood;
Till all night's spark-sprayed dome is stunned with quick airquakes
    of gold,
Precipitous ephemerae and crepitations, streaked
With shivering scars of wounds stabbed by the rays of soaring stars,
Stars piercing scarlet holes, holes bleeding light,
Light strained through silk, silk blobbed with black,
Black blurred with sea-water, blue . . .

3. ON THE GRAND CANAL

The palaces are sombre cliffs by night;
Some pierced with square-hewn caves,
Grottoes where chandeliers like stalactites
Frosted with electricity blaze dangling in the midst
Of sad high-ceilinged salons' tepid haze;
Or semi-concealed by casement shutter-slats
The twilight velvet cloister-cells of lives
Upon whose intimacy we may gaze
As we slide by, nor stir to any flutter
At solitary privacy intruded on
The page-perusing half-glimpsed inmates' eyes.
Others among these wave-lapped marble fortresses
Within which the patrician past lies passively besieged,
Long before midnight look already left unoccupied
Except by somnolent and unseen soldiery,
As from their blank embrasures only blackness
Broods on the glimmering oracle of the tides
That slowly rise and fall about their feet.
One summer night a passenger upon a steamer, I
While we were floating past before them, tried
To read the mystery of the city's palaces
In the framed scenes and silhouettes displayed
To all that sail down the Canal, and when we paused

A minute at a *stazione* raft, looked up and saw
And seized on instantly, a young girl's head
In a near window, her sweet fresh-coloured face
Vividly lit with eagerness, whose aspect made
Me wonder what it was she held before her
And seemed to read from, what the text and page
Of Goldoni or Shakespeare she rehearsed.
But as the steamer stirred again I saw
It was a fan of playing-cards she held,
A lucky hand, as her expression showed . . .
I wished that lovely face good luck in love,
Though my excitement at the glimpse of her
Swiftly became an elegiac feeling
As the boat's motion swept her from my sight.

BIRTH OF A PRINCE

Many of us remember, too, how very young
And unlike the naïve idea of parents, our own were,
(Though many also may have been less fortunate), when we
Proudly were brought by them into a world of care—
Such genuine gentle care and such brave faith
In the great future which they knew that we should see.
Many also were born within sound of the wind
That can blow no man good, the howling wind of war,
National adversity and Winter. In the historic park
A horn like Herne's was heard; the times were dark;
And the great royal oak creaked in the blast
With grief, its branches cracking, though unshakable it stood.
Another daybreak, and behold with dripping boughs
Uprise after that storm a tree that stands because it stands
For true Peace rooted in the right, from which no wind that blows
Shall shake the many birds whose song is still heard in these lands.
No bird but very bat is he who cannot see
A smile best recognized in solitude
In this momentous birth, nor hear another tongue
Than that of public oratory still speaking through the roar
Of loyal multitudes, asking God grant that we
Give birth to the world's only Prince, *Puer Aeternus*, He
Whose swordlike Word comes not to bring us peace but war
Within forever against falsehood and all fratricidal War.

REX MUNDI

I heard a herald's note announce the coming of a king.

He who came sounding his approach was a small boy;
The household trumpet that he flourished a tin toy.

Then from a bench beneath the boughs that lately Spring
Had hung again with green across the avenue, I rose
To render to the king who came the homage subjects owe.

And as I waited, wondered why it was that such a few
Were standing there with me to see him pass; but understood
As soon as he came into sight, this was a monarch no
Crowds of this world can recognize, to hail him as they should.

He drove past in a carriage that was drawn by a white goat;
King of the world to come where all that shall be now is new,
Calmly he gazed on our pretentious present that is not.

Of morals, classes, business, war, this child
Knew nothing. We were pardoned when he smiled.

If you hear it in the distance, do not scorn the herald's note.

FRAGMENTS TOWARDS A RELIGIO POETAE

'Given that a man has genuine experience of the interior life, then let him boldly drop all
outward disciplines, even those practices which thou art vowed to and from which neither
pope nor prelate can release thee.'

MEISTER ECKHART

I

The Son of Man is in revolt
Against the god of men.
The Son of God who took the fault
Of men away from them
To lay it in himself on God,
Has nowhere now to lay God's head
But in the heart of human solitude.

2

The way to Life is through the entrance into Night:
The recognition of the Night wherein each man
Must have at first existence: knowing not
The Whole, and yet believing that he knows,
And through such blind belief made blind to Truth.
Truth is that Truth must first remain unknown to me:
That in the unknown dark I feel alone.
In this state only can true being wake
To knowledge of itself through consciousness
Of the non-entity that it is born from and of the desire
For Being, Truth and Light and Human Day.

3

Dear Nameless God, must I say Thee
When I address you? or should I now try
When speaking in close intimacy to friends
To call them Thou, and make sincere and true
What has become archaic in a world of falsity?
An overwhelming contradiction rends
Apart all possibility of our addressing You
Until we have within ourselves made one
The will to self-exist and our desire to be:
To be with God, and not pseudo-divine
Scorn-inspired self-deceivers dreading most to be alone.

4

This world remains 'the World',
An empire under rule
Of a confederacy of lone wolf-hearted birds:
Imperial eagles, each unrecognized
Except by his own world.
No self-reliant haughty bird of prey
Can rule the world wearing an Emperor's crown.

The ancient eyrie-world remains grimly convinced
That no society can thrive without 'religion';
And every now and then duly inaugurates
Another mission drive to raise the same old corpse.

That there is Justice in the world
Even the fool who hath said in his heart
There is no God
Would be unlikely wholly to deny:
But if he did, even he would not be such a fool
As the man who declares that there is Justice in the world
And that he can not only see it plainly but must proceed to
     administer it with perfect justice.

There is no perfectly just man
Because the vision of Justice is the pleasure of God alone.
And that is why the divine part in all men
Longs to see justice and to live by it;
While the enemy of God that is in each of us
Is always trying to make us satisfied with what we can see of Justice
     without God,
As though He were bound to ratify automatically
Whatever a man-made judge with his own reason decides is just
Provided a sufficiently large number of other men be persuaded to
     agree with him.

6

There are no harsh laws,
Only laws that in a self-respecting society would be regarded as
     unnecessary.
There are harsh souls and law-encumbered spirits
Who inflict their conception of decency
On men and half-animals and human beings alike;
Who expect our respect
And would not seriously believe it if told we could feel none for
     them.

Really religious people are rarely looked upon as such
By those to whom religion is secretly something unreal;
And those the world regards as extremely religious people
Are generally people to whom the living God will seem at first an
     appalling scandal;
Just as Jesus seemed a dangerously subversive Sabbath-breaker
Whom only uneducated fishermen, tavern talkers and a few blue-
     stockings of dubious morals
Were likely after all to take very seriously,

To the most devoutly religious people in Jerusalem in Jesus's day.
Let the dead continue to bury the dead as they did then,
And let the living dead awaken and greet with joy the ever-living.

8

Always, wherever, whatever, however,
When I am able to resist
For once the constant pressure of the failure to exist,
Let me remember
That truly to be man is to be man aware of Thee
And unafraid to be. So help me God.

9

Christ was hung up to die between two thieves;
    And much mirth did the spectacle arouse
    Among the populace who'd heard Him say
    That He was One with God and their true King:
    Look at Him now! It's strange that God allows
    His Son to come to grief like that, they cried;
    All such pretentious scoundrels end that way!
    God's Son! Whoever heard of such a thing?
    There hangs our King, a thief on either side!

For Christ was executed by the general will,
    Officially and popularly execrated, thrust
    Out of this life in ignominy, put
    To death outside the righteous City's wall:
    An unsuccessful outlaw and a grim warning to all
    Who would disturb Pax Romana with thought,
    With the unmanly doctrine that all men
    Should love fraternally their fellow man
    Instead of warrior-like despising him.

10

Though towards the suburbs the city becomes wan
And dark with the weariness of the women who have to queue
Outside the horse-butcher's or for the home-bound bus,
On even the busiest days the sun sometimes paints propaganda
For the possibility of the Kingdom of Heaven on earth
Over the prices scrawled in white on the shops' plate-glass,
And the attic window-boxes above the market
Offer tribute of happy beauty to the omniscient Heavenly Eye.

THE SECOND COMING

In the dream theatre, my seat was on the balcony, and the auditorium had been partly converted into an extension of the stage. Several little Italia Conti girls ran forward past my seat from somewhere behind me, and one of them clambered over a ledge and seemed to fall (she must have been suspended by a wire) to the floor below. She gave a small scream: 'God is born!' On a little nest of straw on the ground close to where she had fallen, a baby doll suddenly appeared. At the same moment, a hideous scarecrow-like Svengali-Rasputin figure, mask larger than life-size and painted rather like an evil clown in a Chagall apocalypse, playing an enormous violin which somehow contrived also to suggest the scythe of Father Time, rose upon the circular dais in the centre of the auditorium. I realized at once that he was the personification of Sin and Death. 'When I play my tune, there is not a single one of you all who does not join the dance!' I was most painfully moved by the strident yet cajoling music and by the knowledge that what he had said was nothing less than the truth. Everything then began to move around confusingly. On the darkened stage, thick black gauze curtains had lifted, and one saw a squat black cross outlined against a streak of haggard white storm light across the back-cloth sky. Finally, the stage was full of menacing, jerkily swaying bogies, thick black distorted crucifixes with white slit eyes, covered with newspaper propaganda headlines, advancing towards the audience like a ju-ju ceremonial dance of medicine men. At the very end of the performance, a clearly ringing voice, representing the light which must increasingly prevail against these figures, cried: 'All propaganda that is not true Christian revolutionary propaganda is sickness and falsehood!'

A LITTLE ZODIAK FOR KATHLEEN RAINE

ARIES

Augustly awe-inspiring creature, whose famed Fleece
And cornucopiae-like Mosaic Horns of gold
Foreglimmer from afar the Great Year's harvest of pure peace;

Entangled in the thicket of the World Roof-Tree's dense leaves,
Immortal Ram, like Absalom dangling his slain youth's gold
Caught on an oak bough in the wood, for whom the Father grieves:

Suspended is your splendour in the domed space of the dark,
O scion of the sacred flock, in scripture spelt of gold
The legend of your leap ever recorded in mid-arc.

GEMINI

Each looks towards his brother and sees yet one more than him;
In friendship with each other sealed, they both remain unmet.
Their eyes still gaze towards the misty heights that precede Time;
Whatever one of them looks on, the other will forget.

TAURUS

Lunging Beast,
Bulging hide,
Fatalist
Ruby-eyed,
In coiled maze
Or sordid ring
Blood betrays
Butcher King.

CANCER

This fishy thing that sideways crawls
    But neither swims nor flies,
Elects to dwell in shellac walls
    And has protruding eyes.

About this sign I've nothing more to say.
I'm not born in or near it anyway.

LEO

No smaller than the Sun amidst the mid-day sky,
With oriflamme-spiked ruff of red mane stands
    This calm carnivorous King
    On tufted turf among
The gentle field-flowers of his wild domain;
    And brands
    With tawny patch of scorch
The green herbaceous velvet ground on which
The leonine supremacy is thus embroidered plain.

VIRGO

Where waterfalls and willows and interstices
   Of nightblue undissolved by day perform
The offices of backcloth and of trellises
   For briars in bloom to climb upon and swarm
      With emblems white and red
      About her uncoiffed head,
A young lady sequestered and immaculate,
Scarce asking whether any less hermetic state
      Await her, may be seen
      Plaiting a garland green
For Chastity to wear when she is dead.

LIBRA

   O unjust man behold
   How she must stand blindfold
   Who personates the word
   Justice, and in one hand
   Wield naked sword as wand
   Who with the other lets
   Two equidistant plates
   Dangle, while she forgets
   Which yours is, which your fate's.

SCORPIO

   Here is a beastly jewel!
   Its tail can cause to groan.
   If scorned or feared it will
   Lurk under every stone
On the wide avenue towards Success
That seems to lead out of the wilderness.

SAGITTARIUS

I, Father, with my little Bow
Plant my munitions high and low;
Trusting, should they shoot up by night
The buried dragon will not bite.

CAPRICORN

Alone alike elect on heights of prophecy
   And exiled on the darkling plain of Chance,
Trailing the guilt that makes worlds wildernesses, he
   Performs his tragi-comic limping dance.

AQUARIUS

This burly bent, much burdened figure, who
Is he, I wonder, and what does he do?
Old Atlas, is it, staunchly straining still?
Atlas? Oh, no. This man's about to spill
Into some hole from his pot lots of sea.
Of sea? I see.—Unless it's Hippocrene.—
But it's not pink, I think, as that would be:
Perhaps it's just plain drinking-water?—Yes,
That probably would be the wisest guess.

PISCES

They glitter, but they sing
   Seldom; rather than swim
They slide through that thick element the waves
Roof in; swing the slow loop
Of a lassoo through which
In reflex they can swoop
And thus with cunning catch
In their own track themselves. And then they sweep
   Down sheerest slopes
   And swerve
   Round sharpest curves
And leap abruptly up, like swift sea-larks,
To burst through their sky's rolling clouds of foam
And briefly warble, before sinking home,
A stave of bubble-song; to which no sailor harks.

AFTER TWENTY SPRINGS

How vehemently and with what primavernal fire
Has there been voiced the seasonal conviction that new birth,
Aurora, revolution, resurrection from the dead,
Palingenesia, was about to be, was near,
Must surely come. Of course it shall, it must.
The bones shall live, the dust awake and sing.
I hope and trust I shall be there. But seriously,
If it has not already come, and it is we
Who lack the faith to recognize it, if the sun
That shone upon the just and unjust does not shine
This spring upon the risen dead, then what a long
Business this getting born again must be. We dead
Are living, really; and the living are asleep,
Lawrence; and gladly in their sleep they read
The Twentieth Anniversary reprint of your writings, stirred
Fitfully for a while to more impassioned dream.
For many love you now, Redbeard, and wish you had not died
In bitterness, before your time. On dead man's isle,
We who survived you and are struggling still today
(If very feebly and unostentatiously)
For life, more life, new life, fine warm full-blooded life,
Are reconciled with patience, on commemorating you.

ELEGIAC IMPROVISATION ON THE DEATH OF
PAUL ELUARD

A tender mouth a sceptical shy mouth
A firm fastidious slender mouth
A Gallic mouth an asymmetrical mouth

He opened his mouth he spoke without hesitation
He sat down and wrote as he spoke without changing a word
And the words that he wrote still continue to speak with his mouth:

Warmly and urgently
Simply, convincingly
Gently and movingly
Softly, sincerely
Clearly, caressingly

Bitterly, painfully
Pensively, stumblingly
Brokenly, heartbreakingly
Uninterruptedly
In clandestinity
In anguish, in arms and in anger,
In passion, in Paris, in person
In partisanship, as the poet
Of France's Resistance, the spokesman
Of unconquerable free fraternity.

And now his printed words all add up to a sum total
And it can be stated he wrote just so many poems
And the commentators like undertakers take over
The task of annotating his complete collected works.
Yet the discursivity of the void
Diverts and regales the whole void then re-enters the void
While every printed page is a swinging door
Through which one can pass in either of two directions
On one's way towards oblivion
And from the blackness looming through the doorway
The burning bush of hyperconsciousness
Can fill the vacuum abhorred by human nature
And magic images flower from the poet's speech
He said, 'There is nothing that I regret,
I still advance,' and he advances
He passes us Hyperion passes on
Prismatic presence
A light broken up into colours whose rays pass from him
To friends in solitude, leaves of as many branches
As a single and solid solitary trunk has roots
Just as so many sensitive lines cross each separate leaf
On each of the far-reaching branches of sympathy's tree
Now the light of the prism has flashed like a bird down the dark-
      blue grove
At the end of which mountains of shadow pile up beyond sight
Oh radiant prism
A wing has been torn and its feathers drift scattered by flight.

Yet still from the dark through the door shines the poet's mouth
      speaking
In rain as in fine weather
The climate of his speaking
Is silence, calm and sunshine,

Sublime cloudburst and downpour,
The changing wind that breaks out blows away
All words—wind that is mystery
Wind of the secret spirit
That breaks up words' blind weather
With radiant breath of Logos
When silence is a falsehood
And all things no more named
Like stones flung into emptiness
Fall down through bad eternity
All things fall out and drop down, fall away
If no sincere mouth speaks
To recreate the world
Alone in the world it may be
The only candid mouth
Truth's sole remaining witness
Disinterested, distinct, undespairing mouth
'Inspiring mouth still more than a mouth inspired'
Speaking still in all weathers
Speaking to all those present
As he speaks to us here at present
Speaks to the man at the bar and the girl on the staircase
The flowerseller, the newspaper woman, the student
The foreign lady wearing a shawl in the faubourg garden
The boy with a bucket cleaning the office windows
The friendly fellow in charge of the petrol station
The sensitive cynical officer thwarting description
Like the well-informed middle-class man who prefers to remain
    undescribed
And the unhappy middle-aged woman who still hopes and cannot
    be labelled
The youth who's rejected all words that could ever be spoken
To conceal and corrupt where they ought to reveal what they name.

The truth that lives eternally is told in time
The laughing beasts the landscape of delight
The sensuality of noon the tranquil midnight
The vital fountains the heroic statues
The barque of youth departing for Cytheria
The ruined temples and the blood of sunset
The banks of amaranth the bower of ivy
The storms of spring and autumn's calm are Now
Absence is only of all that is not Now
And all that is true is and is here Now

The flowers the fruit the green fields and the snow's field
The serpent dance of the silver ripples of dawn
The shimmering breasts the tender hands are present
The open window looks out on the realm of Now
Whose vistas glisten with leaves and immaculate clouds
And Now all beings are seen to become more wonderful
More radiant more intense and are now more naked
And more awake and in love and in need of love
Life dreamed is now life lived, unlived life realized
The lucid moment, the lifetime's understanding
Become reconciled and at last surpassed by Now
Words spoken by one man awake in a sleeping crowd
Remain with their unique vibration's still breathing enigma
When the crowd has dispersed and the poet who spoke has gone
    home.

PAUL ÉLUARD has come back to his home the world.

SENTIMENTAL COLLOQUY

Daphne: The evening in the towns when Summer's over
        Has always this infectious sadness, Conrad;
        And when we walk together after rain
        As darkness gathers in the public gardens,
        There is such hopelessness about the leaves
        That now lie strewn in heaps along each side
        Of the wet asphalt paths, that as we turn
        Back to the gardens' closing gates, we two,
        Though in our early twenties still, seem elderly,
        Both of us, Conrad, quietly quite resigned
        And humbled into silence by the Fall . . .

Conrad: My dear, even your Mother is not elderly!
        A woman is a girl or an old maid.
        Yet I too do feel muted by this twilight;
        For as it ever is the tendency
        Of dusk to fall, and of past Summer's leaves,
        At this time not of day but of the year,
        To drop from trees, so surely must we fall
        Silent if we take lovers' strolls in Autumn
        Hoping we'll not fall out before the Spring.

Daphne: I hate you, Conrad, if that's what you're hoping!
I don't believe you think I'm a 'young girl'.
There is already in the air that hint of death
That when we breathe it makes us winter-wise.

Conrad: I do not think we to ourselves appear
A pair of fledglings. Let the middle-aged
Be sentimentally aware of their maturity
But let us not seem to invite their envy.
We shall be like them sooner than we think.

Daphne: There go a couple really bent with care:
Oh, look! how they both love each other, though,
In spite of—

Conrad: Why, you only speak your wish,
Daphne, you've not looked close enough!
A pair of ancient fish, my love, out of the deep:
Mute and expressionless they loom and pass
On their dim way across the ocean floor
Of roaring London.

Daphne: Conrad, how long ago
Did we sink drowned in it? Little you care
For two such poor old phantoms. Sink or swim,
We have no choice, since gravity descends
And we although our love's still young
And though true love's immortal, are as old
And sink as fast as hearts of stone, if we pretend
We care for no one but ourselves,
Failing to recognize that that's who they are.

Conrad: You will become a Sybil, sweetheart, soon.

VIII
LIGHT VERSE

AN UNSAGACIOUS ANIMAL
or THE TRIUMPH OF ART OVER NATURE

The Master of *The Monarch of the Glen*
Was making once a sojourn 'neath the roof
Of an admiring Peer, Lord Rivers, when
Occasion rose which put to sternest proof
That intrepidity and tact which had
Secured for him familiar intercourse
With Nature's greatest gentlemen and made
Him reverenced alike by man and horse.
For while his fellow guests one afternoon
Were raptly gleaning Landseer's dicta, sound
Of lawless canine truculence, which soon
Became intolerable, made him pound
With sudden fist the tea table and cry:
'What insolence of importuning cur,
What rumour as of kennel mutiny
Is this? Shall Man the Master then defer
To a hound's ill-bred fury? Follow me:
Let's to the stable-yard whence these barks come,
And I will prove to you that Art can be
A force more sure than blows to make dogs dumb.
I who not seldom with forbidding gaze
Have known how to persuade huge Highland kine
To emulate the Southern cow's sweet ways
And made whole shaggy herds hang on the line,
Will there, if it amuse you, demonstrate
A sovereign power yet stronger than the eye's:
That of the Human Voice, which is so great
That it can Lions strike dumb with surprise!'
Some of the painter's intimates had been
Already privileged to hear his skill
In imitation of the less obscene
Sounds with which animals are wont to fill
The atmosphere of jungle, swamp and glade
When moved by meal-time longings or by bliss
To self-expression. For some years he'd made
The feat his study, and could bellow, hiss,
Roar, bark, snarl, with a realism which
Was quite astonishing, till in no part
Of all Victoria's realms was known so rich
A repertoire of Imitative Art

As that perfected by the great RA.
In view of this, it hardly will seem queer
To any that all present there that day
Excitedly accompanied Landseer
Out to the stables, craning and agog,
To watch him stride, masterfully serene,
Towards the kennel out of which the dog
Surveyed defiantly the crowded scene
With jaws aslaver and keen fangs exposed.
Then, not without surprise, they saw him fall
Down on his knees! It was by some supposed
This was in order piously to call
On Providence for aid; but they were wrong.
His aim was to confront the renegade
As man to man (or—dog to dog?). Ere long
That wretched animal's vile din was made
To seem the fretful yap of Pekinese
By an appallingly hyenine bark
Which evidently made the dog's blood freeze,
For his rebellion ceased at once, and stark
Terror replaced the murder in his eye.
The artful mimicry of Landseer proved
So awful that the beast which recently
Had rivalled Cerberus himself, now moved
With such violence away from the advance
Of the superior barker, that his chain
Snapped, and he crossed the yard swift as a glance,
Leaped o'er the wall, and never was again
Seen anywhere on Lord Rivers' estate.
Landseer, on rising, found that only one
Of those who'd watched him still remained to fête
His triumph. 'Twas his host, who breathed: 'What fun!
How good of you to teach them how, dear old
Dog-lover! But come now, your tea's quite cold.'

LE DÉJEUNER SUR L'HERBE: A PASTORAL

LA BELLE-DAME-SANS-MERCERIE:
Thank goodness, *mes chers amis*, that you do not
Object to *negligée*. I was far too hot.
It's such a pleasure to find now and then
Friends who do not just look like gentlemen.

LE DUC DE PROFIL:
Dear *Dame*, we are most flattered by the candour
It pleases you to show towards us and our
Best feelings, I assure you, are excited
By this display of favour. I'm delighted
To find it was not a mistaken hunch
I had about you. Now let's have our lunch.

LE COMTE D'À CÔTÉ:
Let's hope no herd of any kind will pass.
Not only cows, you know, lunch upon grass.

LA DEMOISELLE AU FOND DE LA TOILE:
I should imagine that huge meal they've had'll
Make them too sleepy much to want to paddle.

THE DECAY OF DECENCY

When Man becomes what he calls 'adult' and stops taking himself
    too seriously
He is apt to get oddly pedantic about the proprieties while even
    more loose-mouthed than ever
Should anyone foolishly tarnish their honour by calling his into
    dispute.
A jealous prude, this much (though, alas! unavailingly) chipped
    blockhead
Is always able to preserve his dignity and our decorum in even the
    most embarrassing situations
Provided the common law still allow him a remnant of the old
    apron-cloth with which to camouflage his flyblown shame.
Proud resident of no mean or indecent city, he can be trusted
    always to think of shielding his neighbour
Who may although bald be still nevertheless immature,
From the horror of being debauched by an unwonted glimpse
Of the unavoidably material source of the carnal life whose too
    often notorious facts
Are not, when unvarnished, the kind one can mention in public, in
    verse or to even, of course, one's own wife,
So deep-rooted, apparently, is our innate sense of reverence for
    whatever must never be spoken of
Till all representatives of the gallant little Sex the Serpent seduced
    have withdrawn to another room,

173

Or as resoundingly and full-bloodedly as you like so long as it's only
    in good clean working-class fun!
Should some contradiction begin to appear in the bourgeois
    conventions, it can be explained
As due to the obstinate whims of our animal nature, but nowadays
    not, if you please,
By referring to what we no longer regard as a dogma, our dreadfully
    common first parents' Fall.
Take away a man's responsibility to our self-respect, and you may
    rob him of half his charm.

WITH A CORNET OF WINKLES

(VERS DE CIRCONSTANCE)*

O bravo! For a maladif, mandarin-miened, mauve melody-man
With a glittering, lissome, pat-prattling lute—
Que c'est beau! as he lo! hums and haws,

And soon again haws, then heigh-ho! how he hums
And whilom most becomingly strums
On his poignantly Quince-flavoured lute!

Ah! what is it, the sensitive rattling of lute-players? What
Is it that makes them persist undeterred by the glees
Of the slums' glum goloshes-clad mummers

Who in average summers become far more raucous than we
Ever dared (my shy mandolin!) be: for much cow-heel and hot
Cake—nothing rare, just plain fare—ever wait

For glee-singers to guzzle as soon as they're through
With their dreary commercial drum-dub rub-a-dub (there's the
    rub!)
But in Hartford all's mum for a mo

I mean moment: as finnicking wryly with *curedent* that one
Still remaining Romanticist who's fully weaned of such pap
(If from involute vocables glibly redundant mayhap

* This tribute to Wallace Stevens was written before he won the
Bolingen Prize in 1950.

Not so drastically weaned as he might be!), the Great
Gabbo, sole plum among lute-players left
To preserve some bravura or any finesse and who yet

Would never have dreamed of abandoning strumming unless
He'd been vexed (as he has) by the pip of some fruit
Among molars illicitly lingering,—ceases

Perforce to oblige with the teetering tittery-tum
Ti-tumtum titivating of his lacquered lute:
Willy-nilly sit silent. Is not the hush rum?

Yet withal scarcely *more* rum,—and rum quite a lot
Will agree it does seem,—than that one should these rum rumin-
Ations style (aptly? or do you think otherwise? Ah!) philosophical:

Though that's what this old-master lute-master opines
That his airy (Hurray! Tipperary for ever!), his endlessly
Varied *echt*-lyrical lute-ditties are! 'Tra-la-la!'

One might here interject, without being inept. I had not,
I confess, ever fully imagined how rum they might be
(Ruminations), if once but a maladif, canny, a hey-de-ho whole-

Heartedly cogitatorial *maître-de-luth*
Were to get *un vrai goût* for Philosophy. What as we but
A brief moment ago were about to enquire: What the Hell, O but
    What

Are the, or rather, what *is* (put it plainly) the point
Of this perfectly awful attempt at a parody, piffling without
Doubt to the point of provoking a petulant pout

Of disdainful demur (a mere *moue*, that will do)?
Tootle-loo, Tilly-loo, for no earthly *Poetic What's-What & Who's-
Who* will now ever again deign so much as to look at these too

Caca-caca-cacophonous doodlings of mine
And not think them impudent snook-cocking. Candidly, look you,
    what do
People put up so patiently with them for? Answer: *Nein, nein,*

Far from it, they don't, don't deceive yourself; and what is MORE
'The Corn's Blue!'

175

THREE CABARET SONGS

1 A BRIEF BALLAD OF THE PARALYSED UPPER LIP

Oh, the Bland Maid of Kensington,
  She Lived there in Sin
With a Bluff Ex-Young Ladies' Man
  With a Permanent Grin.

Oh, the Life that he Led her there
  Was Well-dressed, but so Slow
That she Longed to Abandon him
  Yet could never quite Let go.

Oh, he Grinned at her Frailty,
  She smiled at his Pride!
And Long though this lasted
  They neither of them Died

But Continued in Kensington
  To Adorn every Day
The Saloons of Three Locals
  Well-known to be Gay,

Where her Blandness, his Bluffness
  Continued to the End
To Convince all Acquaintances
  That She was his Friend!

Oh, what really is Dreary
  About all such Brave Pairs
Is that Being Godforsaken
  Is the Least of their Cares.

2 WHAT A WAY TO WALK INTO MY PARLOUR, LITTLE MAN!

Ere the hour for aperitif's over
  Take care your tongue doesn't work too loose,
Though if you want wit, you're in clover!
  This young woman, you see, is no blue goose
Nor green-stocking, by Jove!

If you're looking for someone to lie to
    About all the fictitious feelings
You feel you should feel, my reply to
    Your ogling is, I have no dealings
With people unreal.

Keep your labels for people who need them;
    I cannot be pigeonholed neatly.
As for your ideals, I exceed them
    So far, I surpass them completely
Please beat a retreat.

Let me tell you that all this persistence
    Is worse than absurd, and I must say
What I see of your mode of existence
    Inspires in me only disgust. Lay
Off, Less-than-the-dust!

3 SIZZLING SECLUSION: RUMBA

Don't murmur or mut-
Ter: No, no! or: Tut-tut!
I'm deep in the rut
    Your scent rouses.
It's too hot for a hut
Or a bungalow but
I've had such a glut
    of pent-houses.

I could just make do
With a wigwam for two
Though I'm not going to stew
    In a leather one:
One of muslin instead,
Draped around a deep bed,
    A soft feather one . . .

THREE VERBAL SONATINAS

I HISPANIC

To Rafael Nadal

This is for reading
One wild long winter night
Before the long and dreary
Wild winds have come to fright
Us all with their emotion-
al and spicy Winter's tales,
Have come and gone and broken
With the force of heavy gales,
Leaving spume upon the mantel
And ice upon the floor—
Where are we now? Good gracious,
It's only half-past four!

If you want to eat them,
These words will always do,
For that's the way to treat them,
And wild Cassandra's too!
There is no time to sweep up
The tears upon the floor,
So crystalline and icy,
Since 'that was half-past four!'

For since that long-past cry went
Up into empty space
The birds have come, and by went
The thrilling parrot race!
O human, all too human,
Are those they left behind,
Baboons are we, to some of us,
And bloody pansies, mind!
But Christmas comes but once a year,
Can Spring be far behind?
I've lost my rhyme, and metre too,
Whatever shall I do?
The time has come to wipe us up
Again, thou Winter Wind!

O thrilling Sound, the Winter round
'We have had Spanish Flu!'
And when the wind-up wind does sound
T'will come again for you!
The pen, the pen, the bright blue pen,
It has turned up again!
So now we'll write till candle-light
Has drenched the world, the Main
(The Spanish One), and all,
Oh, all the World's in sight!

2 NEO-CLASSICAL

To Terry Clare

This is a strict
And very clean song,
Not without words but,
Not to be *too* long,
Without a tune,
Unless you like to
Make up your own
('Ere the clock strikes,two!)

It's very brief,
(A Sonatina
Has to be!) and rolled-
Up like a concertina,
The basic row,
The saying goes
That life is grief
Without a song, so
Anything goes!

A Spanish song's
No more in order
Since 'Half-past four!'
Came in and roared a
Great rallying cry,
The other evening,
Another time, another . . .
(What rhymes with 'evening?')

This, as I said,
Is meant to be
A poem read,
No symphony!
Don't try to sing
But only listen
A bit to that,
And then to this'n

We'll soon be through!
I think so, don't you?
Be through 'ere two,
Eh? Will you, won't you?
Well, we shall see
As soon as sight comes
To light the burden,
And some light hums

May see us through, till
This opus ends, as
It opened, with a
Nice strict (has
It been that?) tune
In words, so
That we will soon
Reach the last stanza.

(What rhymes with that? Oh,
It doesn't matter!) Then
If not too flat, Oh,
We will have done well,
And eight lines each
A stanza helped
Us to at last reach

This final one!
It's been brisk going.
But we've had fun!
Such Ah- and oh-ing,
Have done at last,
And well before two!
Just not too fast!
Now we are right through!

3 · SERIAL

To Humphrey Searle

This starts with ATO
And then goes on to be
A set
Of variations. Of variations
On ATO. These three
Letters comprise
The basic row,
Like ABC. Instead of Twelve
Notes, we have three basic
Letters. Compare
These three
With their associations,
As, for instance, Ash,
And Tin and Oranges. And then
Shh. Wait for the next,
The next world coming on,
This earth to change: W.R. & B.
Nice to be good
Nice to be rich
Nice to be full
And beach

As I was saying,
Before I was
Politely interrupted, take
Verbal associations, such as
This time,
Ants, Offal, Tunes.
On some such tunes as these
All variations are composed.

Next Variation: Take
Three Basic Rows, such as
We have above. Then choose
Some completely new associations
This time. Take Orange,
Antiques and Turpentine.
These form an as it were
Complete
Contrast with the previous

Variation. (At least I hope
So). Then,
After such an ATO,
Atonal kind
Of variation, we pass on
To the next variation which
Is this time once again
On OTA,
So that we have
By this time come
Back to the basic
Triad: O for Oxford,
T for Times, A
For Alphabet.

But what comes next?
Let's see.
The ABC. I see.
C is the Basic Note,
C major, that is,
For all atonalists
Who know their Latin. Let's
Conceive that A is B, and B
Is C. Where does
That get us?
Cannibals, Beach-girls.

And Anannapolis.
This will be Greek to those
Who think that Prose
Is Simple Gospel. Now I think
That Variation's
Done with. What comes next?
Next comes a further,
A yet further variation,
This time on OAT
(I do not mean the cereal)
Now, oats are sown,
But that's not Serial.
And so we're back
To Purely Verbal
Associations, this time: Octaves,
Apples and Tintype.
That should be plain,

Quite plain and simple
As quite befits
This plain and simple set
Of Variations.

So to complete
This complex set
Of simple variations, let
Your mind next dwell
On OAL, just to make one
Small unexpected and
Discordant letter creep
Into the Game. With L
Let us associate a single thing,
Say Lint, or Lemon, Light
Or Loganberry. This
With O and A makes up
A sort of supersidiary
Penultimate
And temporary variation,
LOA, loa, as in Loaves,
Though, unfortunately, we
Can't work in Fishes
So let's get back
And let us see,
Once more, just how
To bring this Sonatina (Verbal)
To a conclusion fit
For such a thing. Suppose
We now again evoke
The Basic Three. No?
That would be rather dull? Well
Let's try a free
And definitely final this time
Association. Say O for Oak,
and T for Tank, and A
For Arcady. OK. Let's go.

Well, that went quick! Now let's
Go slow, a little. Then
Will come the turn
For ATO to grow
Once more familiar, like a rare
And convoluted tune;

And soon
Will come the Twin
For Ashes, Oats and Trees
To come at last
To their last Resting Place.
An English man here says
That English men are
Liars: Is he reliable?
That is the final, unexpected
Intermitted question that makes this
Still a surprising variation.
If only LIA
Had crept in earlier, we could
Account for it. But no,
Mere wordplay cannot lie.
So let us say goodbye
To ATO and ABC, and
LOA and all
Their variants, and end
Not with a dying Fall, but with
The Final, First and Tonic C,
True 'Art of Harmony'!

IX

NIGHT THOUGHTS
Radiophonic Poem

(1955)

NIGHT THOUGHTS

> Aber weh! es wandelt in Nacht, es wohnt, wie in Orcus Ohne
> Goettliches unser Geschlecht . . .
>
> HÖLDERLIN

But alas! our generation walks in night, dwells as in Hades, without the Divine . . .

I THE NIGHTWATCHERS

[*Voice A*]

Let those who hear this voice become aware
The sun has set. O night-time listeners,
You sit in lighted rooms marooned by darkness,
And through dark ether comes a voice to bid you
All be reminded that the night surrounds you.

[*Voice B*]

Around us, as within us, battle rages.
Enveloped in obscurity, our enemy,
An emissary from the world of shadows,
Assails us from an unknown vantage-point,
Observes us unawares, usurps initiative
And uses it to inspire such distrust in us
That we must now suspect him everywhere.

[*Voice C*]

Let those who hear my voice become aware
That Night has fallen. We are in the dark.
I do not see you, but in my mind's eye
You sit in lighted rooms marooned by darkness.
My message is sent out upon the waves
Of a black boundless sea to where you drift,
Each in a separate lit room, as though on rafts,
Survivors of the great lost ship, *The Day*.

[*Voice A*]

Let those who hear our voices be aware
That Night now reigns on earth. Nocturnal listeners,
The time you hear me in is one of darkness,
And round us, as within us, battle rages.

[*Voice B*]

A war goes on within against the shadows.

[Voice D]

Who speaks tonight of war and battle? Go to bed!

[Voice E]

The war? What war? We've had too many wars.
The last War's over.

[Voice F]

                    Go to sleep. Put out
That light. The War is over now. It's late.
Why don't those people go to bed?

[Voice G]

                         Why must we hear
Night-voices always arguing about the state
The world's in? Why can't they forget about it?

[Voice E]

                              War?
Why must we always worry about that? Make them put out
Those lights.

[Voice F]

I'm O so sleepy . . . Now let's talk no more.

[Voice B]

The plane-trees in the court outside my window
Suspend their leaves between me and the street-lamp
That burns all night beside the entrance-arch;
And when the night-wind sets their branches waving
The shadows drift in tattered velvet bunches,
Thick-tangled rags of shadow are set swaying,
That dance like the black flames of a cold bonfire,
Leap up and are cast writhing on my bed.

[Voice C]

Anxiety and dream assail the watchman
Who waits in solitude for night to pass,
And shadowy multitudes with muffled tread
March menacingly round about the vigilant.

[*Voice A*]

'Anxiety and dream,' the watchman said,
'A shadowy tumult that I cannot quell,
Stir round me like a wind through sleeping grass.'

[*Voice B*]

I cannot sleep. These nights are terrible. Yet there is now
Nothing more terrible to be afraid of: We have won
The worst; now we need fear no more, nor hide
Our disbelief in anyone.

[*Voice D*]

Can you believe,
O foreigner I'm thinking of, woman unknown to me,
Lying awake somewhere in Europe, can you now
Believe that you have friends lying alone,
In darkness, overseas, who can imagine how you feel
And wish, and wish—ah, what? What can be done
For anyone, what can we do alone, alas, how can
The lonely people without power, who hardly know
How best to help neighbours they know, help those
Who surely would be neighbours like themselves, if they but knew
How to break through the silence and the noise and the great night
Of all that is unknown to us, that weighs down in between
One lonely human being and another? Who can hear
My thoughts, or know how my heart grieves, or feel
That I just like themselves long to believe
That lonely human beings love each other?

[*Voice E*]

I believe
There's bound to be another war one day.

[*Voice F*]

You can't believe
Everything that the papers say.

[*Voice C*]

Russia, the U.S.A.,
Atomic Power, Foreign Powers . . .

189

[Voice F]

> Go to sleep. Put out
That light! The War is over now. It's late.
Why don't those people go to bed?

[Voice E]

> They're all alike
Those foreigners, you can't trust them, can you?

(Confused Grumbling Voices Fade Out)

[Narration One]

The Tyrant Negativity has usurped power and thrown
Men's captive souls into the silent pit
Of self-confounded Subjectivity.
Immortal souls that know themselves to be
Immortal souls have wings.
But in that pit
All doubt-blindfolded souls must fall like stones—
Fall down without the power to cry out
Unless inspired by Anguish.

[Narration Two]

A stone that falls feels nothing, has no fear
And knows no need, and cannot cry.
A falling stone is not a fallen soul.

[First Mortal Soul]

Now Man benighted huddles in his cave,
In mighty ignorance of what he is and what he's not:
Cave-night which every night
His all-aloneness drives him back into:
This is the dark, familiar, fearful place
Where once again flung down I fallen lie!
Oh! could I but release from far within
My own benighted selfish inmost dark, from deep within
The ever unknown part of me, could I release
One long, long harsh heartbreaking broken cry
That would for once express all that the night
Awakens in me, all that words betray,
Being too flimsy and approximate and too
Precise: could I unsay

190

All I have clumsily but half-expressed, O could I howl
Instead the protestation of my impotence against
The dull omnipotence of stifling soundlessness,
Dull swelling vacancy, that from all sides
Drives with the pressure of incessant passing time
Inwards on me, thrusting me back into the lapsed
Being become non-being where annihilation waits
To swallow all that I have ever been,
Then might I sleep like one whom his own soul no longer hates.

[*Narration Two*]

The cry of mortal anguish from the soul's dark night
Reaches you now, if you will hear it. I will ask
Myself whether you hearing it, if you were God
Would pay no heed but turn away your ear.
You have heard one, but there are countless cries.

[*Second Mortal Soul*]

Shut up, shut off that hateful voice.
Shut up, shut out the Night.
I do not want
To sense the world's obscure plain spread
Out under empty heaven, or to know
That we lost in obscurity are stranded on a sphere
Of earth that spins amidst infinity
Among unnumbered galaxies of spinning spheres
Dispersed in distances so vast that human sight
Swerves backward sickened by the senselessness
Of so much space without a single sign
That consciousness, pinprick adrift in it,
Can seize on to decipher.
Let me be stupefied.

[*Narration One*]

We are always free
To turn away. Our hearts can always harden to refuse
To suffer mortal anguish. There are many anodynes.

[*Third Mortal Soul*]

Drink strength and comfort now out of the well
Of Night, that can so quickly quench our thirst
And as it slowly slakes its own, consumes us all.
The sun sank out of sight and darkness covered us.

I will sit down and close my eyes and wait; sit still and wait,
Though I still somehow cannot yet relax, I feel a weight
Of heaviness that will not let me rest, a load that stirs
And slackens in me, weighing down, wearing away,
With weary will to stay awake when I lie down,
My wish to give up vigil for repose.

[*Nightwatcher's Voice*]

At Night, I often sit an hour out thus,
Attentive to a dull insistent roar—
Or not a roar, rather a kind of cry, and yet
No cry, for that would be a sound too clear,
And what I hear might come from underground,
It is so thick and muffled, and yet hollow-sounding too,
Although not resonant at all, but harsh and dead,
If dead is not too definite a word:
And whatsoever this dull urgent rumour be,
It holds me spellbound by the hour and more,
While I, with a great longing to be free
From doubt about what it can signify,
Gaze up through a small skylight's panes and see
Nothing at all of my star's watch-fire
That may be burning in the black neglected sky;
Do not see even that blank square the window frames—
As though all sight lay blinded in my ears.

And then, returning suddenly again
To consciousness of my immediate self,
I've had a moment's glimpse into the depths
Of solitary absence through which stray
Our tired and restless bodies among all the dead things found
Strewn round them on all sides in an unanimated dream:
Dread has distracted us away from what is here
And what we really are when faithful to the truth;
So we must suffer hopelessly the sullen apathy
That reigns on a deserted theatre's stage
Where all night long we play out our null roles,
In a Morality that could be called '*No Man*'.

[*Second Nightwatcher's Voice*]

I hear a voice that speaks from No-Man's Land
And when just now he said he'd heard a cry
Or some strange sort of sound I thought I recognized

That what I listened to him speaking of I too had heard:
For listen, listen, it begins again! It's the same sound, I'm sure!
On many other nights before I have heard this,
Like sound of distant rioting, that angry voices' sound,
Popular uproar from afar, as though crowds underground
Were pushing upwards boiling to invade the city streets
With hell-hordes hoarsely clammering for blood!
For Blood! For Justice! Bloodshed and Revenge! What cry
Is that I only hear an echo of? Why after all should I
Feel threatened by a thing so far away? Does no one else
Hear what I hear at night?

[*Third Nightwatcher's Voice*]
                    Yes, neighbour, I can hear.
I too have heard those ominous night voices. I hear yours,
You are my neighbour, not a crowd, I'm not afraid of you,
Although I cannot see your face. Then let us not
Mistrust each other, nor be too much disturbed by them.
And do not be afraid of it. If you can hear
The echoes of your own anxiety, if you can bear
To listen to that rumour, then you know at least that dread
Of hearing what you fear has not yet deafened you.

[*Anonymous Mass Voice*]
Fear, fear: you speak of fear.
What is this fear? Is it the fear we dare not fear,
That fear of fear itself, or fear of other's fear,
Such fear as ends
In passionate untruth, self-justifying falsehood without end?
Demonic fear
Of individual guilt, of being caught, of doing wrong,
And fear of failure or of being found a fool,
And fear of anything that might contrast with me
And thus reveal my insufficiency,
My lack, my weakness, my inferiority,
In showing up my difference from itself;
Fear of uncertainty and loss, fear of all change,
Fear of all strangeness and all strangers; and above all else the fear
Of Love, of being loved, of being asked for love,
Of being loved yet knowing one has no love to return;
Fear of forgiveness—
Fear of that love which is so great it can forgive

193

And the exhausting fear of Death and Mystery,
The Mystery of Death, of Life and Death,
The huge appalling Mystery of everything;
And fear of Nothing,
Yes, after all the fear of Nothing really,
Fear of Nothing, Nothing

Fear of Nothing, Nothing, absolutely Nothing.

   [*Voice C*]

Dread of life, and fear of Nothing,
Anxiety and dream assail the vigilant
Watchman who waits in solitude for the Night to pass.

   [*Voice A*]

A blind wind whispers in the sleeping grass.

2 MEGALOMETROPOLITAN CARNIVAL

   [*Voice A*]

When Night has been announced as theme, will it not cause
      surprise
If there is nothing said about the stars? Also it has
Been immemorially the custom to apostrophize the Moon—
In courtly terms, calling her Queen of Night, and to refer
To Cynthia's argent chariot, or some such-like stage-property,
Or improvise some image like that Gallic wit's who saw
The Moon above a steeple like the dot above an I.
Planets and constellations tend to lend themselves to rhapsody,
Having like hosts of lesser stars most ornamental names:
Orion, Mars and Venus, Betelgeuse and all the rest,
That are godsends to poets, shedding lustre on their lines.

   [*Voice B*]

But if I stand tonight,
Not in a poem but in actual fact in, say, Trafalgar Square,
And stare up at the heavens there, what can they mean to me,
The glories of the Zodiac, the lovely names of stars?
Do I see splinters of old myths stuck in the sky above my head?
If stars are visible at all, they're but a sprinkling of pinpricks
Blurred into insignificance by the brilliance on the ground,
Where the City round me celebrates the triumph of the brain
Of man over his darkness, in the effervescent blaze

Of a commerce-sponsored carnival of multicoloured bulbs.
The soot-suffused sky-canopy, shot through with bluish red,
Shuts off from me as surely as do too-familiar names
The mystery of Space.

[*Voice C*]

At night I've often walked on the Embankment of the Thames
And seen the Power Station's brick cliffs dominate the scene
Over on the South Bank, and its twin pairs of giant stacks
Outpouring over London their perpetual offering
Of smoke in heavy swags fit for a sacrificial rite
Propitiating some brute Carthaginian deity;
And thought they stood like symbols for the worship of our age:
The pillars of a temple raised to man-made Power and Light.

[*Voice A*]

And I have sometimes gone out towards midnight
Through streets of dwelling-houses and apartment-blocks
Behind the rows of window-squares of which
Innumerable tired executives prepared for bed,
While past street-corner lamps dogs' pensive escorts
Tugged them on leads along their late patrol;
Through districts full of narrow shady gardens
With strips of black lawn stretching from french windows
To sooty shrubberies, a seedy tree or two,
Laburnum to o'erhang the pavement pilgrim
When summer has transformed these dormitories
By splashing blossom-sprays across their drabness
For a few weeks each year. And have walked on
Until I came out on an open hillside,
A public park space from which one looks down
Upon the mighty Nocturne of the Capital
Whose twinkling panorama's spread below:
Arena sprawling dazed in concrete gloom,
Freckled with sparks and smeared with arc-lights' gleams,
With crawling glares and melancholy glazes,
Slow-sinking monuments and stoic lighthouses:
Mile after mile of tenements and terraces,
League after league of palaces and parks.
Here hover hazes of green sick-ward light,
And there red neon blurrs flick on and off,
In fixed directions avenues stretch sleekly
To disappear in ultimate uncertainty

In regions where the bottom of the sky
Mingles with fumes that rise from the abyss . . .
Fearful and wonderful, that sleepless monster,
Sphinx among cities, Megalometropolis,
Stuns with her grave immensity all eyes beholding her:
One's wonder gapes and quickly palls and falls into dismay,
Knowing the roaring labyrinth deepsunk in Night below
Teems with noctambulists too multitudinous
For any now to fear the Minotaur.

[*Voice D*]

Effulgent filaments in bulging bulbs
Persist in stinging blackness till they've tinged with pallid stain
All wilting areas of opaque obscurity;
Innumerable bulbs that like frost-glazed unpupilled eyes
Pour out incessant bleared lacklustre glare
Upon all public places all night long.

[*Voice E*]

No trace remains in any place of daytime's busy throng.

[*Voice F*]

Behold how every building-block, each bank,
Walls behind which wait bales of ware in yards,
Forums, exchanges, business-houses, stores,
Stand back drawn up behind a film of blankness,
A foreign aspect hazing all façades.

[*Voice E*]

The absent inmates have locked all their doors.

[*Voice D*]

Scarcely a soul is to be seen on any sidewalk at this hour.
Scarcely the word is soul perhaps for such as might be seen.

[*Voice F*]

Their desultory feet move slow and furtively,
Few footsteps far between.

[*Voice E*]

Seeing it now, you'd hardly know the city scene.

[Voice D]

Street-crossing islands stand becalmed; round them no traffic roars.

[Voice E]

All waking feelings now are dimmed, the day-time's passions
   curbed.

[Voice F]

The decent sleep in duty bound. They may emit some snores;
Otherwise they are mute and must by no means be disturbed.
They've made their beds; now they must lie in them.

[Voice D]

They have retired in consequence to do so and are prone.

[Voice F]

Between the sheets, beneath the blankets, parked in cots and bunks,
Stretched out in alcoves, side by side or all alone,
In double-beds or on divans, with lamps out, curtains drawn,
Immobile many millions lie, all interchangeable,
All horizontal humans out of use until next morn.

[Voice E]

No household has been able any longer to refuse
Sleep's standing invitation to its old home castle-keep
There to recline like lords at ease unconscious till next day.

[Voice D]

Everything now has been closed down, shut up and locked away.
The population's breathing is slow regular and deep.

[Voice F]

Although Megalometropolis is unsleeping, night and day,
At times even the city seems to doze off for a spell.
Whether or not it sleeps is hard to tell. I couldn't say.
Brought to a standstill it stands waiting. Empty.

[Narration One]

Enter the Dreams.

The Dreams enter the City.
Drifting in swiftly twisting clouds above the roofs,
Their whirling fever-coloured smoke crosses the moon;
As they race past, its contours blur and tremble.
A moment after, real clouds blot its face.

[Narration One]

Enter the Dream.

[Narration Two]

Enter the Dream's great glimmering park.
Only at first is it still dead of night.
Slide softly, stepping rapidly, at first.
Here there still lingers a strange stealth and stillness.
The beams that fill the early dreams are soft as twilight
In the first place. In this faint light you must move swift as
    swimmers,
Move with short strokes beneath the lowslung boughs,
The grey, long-bearded, overhanging branches
Of ancient trees still lining all these avenues.
You'll have to hurry down these thoroughfares,
Though splendid shops and gardens catch your eye.
All signposts point in only one direction.

[Narration One]

Follow the fingers, you can't lose your way,
It won't take long to reach the central space,
That is the special place you have to find.
Just one street further. Here at last you are.

[Narration Two]

Here is the Circus in the Square that represents
The very heart of the primeval City. Now's the time
To recollect that you've received a secret summons
To a rendezvous with the Unknown, at the foot of the Fountain
That leaps without spray, a thin glimmering quicksilver pillar,
Above the memorial marking the first fatal spot,
The meeting place of the First Person with Persons Unnamed
At the heart of the Forest that grew where the City now stands.

[Narration Three]

The quicksilver Fountain that's hovering there like a column allures
All who enter the lair of the Labyrinth-Omphalos Boss,
Whose domain lies beneath, in the earth. Yet if anyone nears
The Basin too closely, at once it will sink underground.
By the time you've got right to the axis round which the square
    circles,
You will find that it's no longer there.

[Narration One]

Just stand still for a moment. No need to be scared.
Pay no heed to the thunder of traffic, the dazzle of lights
On the walls flashing messages round you on every side.
Soon, just where the Fountain has vanished, the earth at your feet,
At the heart of empirical hubbub, will yawn open wide
And the cavernous Subway's mouth show you the way down inside.

[Narration Two]

Now you follow the steps and descend to the City's true heart,
And are soon in a Plaza illumined more brightly than day
Where more people are hurrying in all directions than up there
    above.
Close at hand is the brisk business district, just under you lie
The platform from which the incessant electric expresses
Go rushing from City to faraway Suburbs, and back from the
    Suburbs again.

[Narration Three]

Here are underground Boulevards bright with Bazaars, here you'll
    find
Vast fields for the shop-window gazer to graze in, Arcades
Branch off on each side, endless Galleries lined with glasscases
    invite
To inspection of carloads of diamondmine loot, of forests of
    flowers,
Tropic fruits piled in tiers, Pin-up waxwork girls posed in parades
To show off new nylons, new sequins, new rhinestones, new lace-
    trimmed furcoats.

[Narration One]

But don't linger too long for a rush-hour approaches and here it's
    unwise

To risk getting caught by the tide of the throng that flows through
    at its height.
Better make your way now to the flights of steps all leading down
To the slow-moving staircases, up to the fast escalators
Descending past columns of spiralling stairs to the level where tubes
Have been bored for the feet to press through from the foot of one
    flight

[*Narration Two*]

Of stepping stones, on to the passages in, then the passages out,
To the thoroughfares out of which more escalators are moving,
    some more
Slowly, the others more quickly, first up and then down, on and on,
On and off, up and up, down down down, go on down, till at last
The wonderful system will crown the true will to success with
    success
As the peace known at zero-hour's peak on the heart of the rusher
    descends.

[*Sleeping Citydweller*]

      Oh! Let me stop, I must sit down!
      I've been deceived, I am confused!
      I must wake from this nightmare soon.
      Among these crowds I've got quite lost—
      Words in the tunnels' roaring drown!

[*Train-Wheels Chorus*]

Hurry up and get on Hurry up and get on Hurry up and get on Hurry
I couldn't care less I couldn't care less I couldn't care less I couldn't
The Main Chance The Main Chance The Main Chance The Main
Get on Care less Get on Care less Get on Care less Get on Care less
Teach a lesson teach a lesson teach a lesson teach a lesson teach a
The Damned are the Damned are the Damned are the Damned are the
The Day of Wrath the Atom Plan the Wrath to Come the Atom
Bomb the Coming Day the Greatest Bang the Biggest Bomb the
Wrath of God the World of Man the Day to Come the Bang the
    Bomb . . . (ad inf.)

[*Guide Voice*]

As you move at a pace that gets constantly faster, your eyes
Are increasingly caught and held fast at each step by one after
Another phrase, slogan and image set up to solicit as much
Of the crowd-individual's attention as each in his hurry can spare.

[Narration One]

You may look where you like for the public's fastidious and only
    permits
Its favourite posters to brighten the walls of such sanctums as these:
Now the principal stations afford a great treat with the constant
    variety
Of the attractions inviting the traveller's mind's eye to rove towards
All sorts of model resorts; at his journey's end wait to stare down
    on him
On his arrival more posters depicting the places abroad he must
Hasten to visit as soon as he can to discover:

[Narration Two]

NEW VISTAS NEW THRESHOLDS NEW PLEASURES NEW BEAUTIES NEW
    BEACHES NEW LIGHT
ON OLD-WORLD INNS NEW WORLDS IN DISGUISE OLD CATHEDRALS
    SPOTLIGHTED
NEW CRUISES TO BEAUTYSPOTS SEA-COASTS BEST SUITED TO NUDES

[Narration Three]

Look! Here posters plaster the best people's eye with huge glimpses
Of Scenes from the Very Best Shows of the Year by the
    Star-Chamber
Critics' Assembly Selected: The Most Highly Praised, the Best
    Advertised, then
The most Noted for Highlypaid Acting, the Most Controversial,
The Brightest, the Loudest, Most Daringly Brutal, and Quite the
    Most Crude.

[Narration One]

The Crowd's hardheaded leaders alone have the leisure to cast a
    glance over them
As they press past down the passage from exit to box-office queue
    but they turn
To present to the next passerby their opinion for what it is worth
    and
He'll then in his turn send it on to be sent on till common consent
Has agreed that it's fit to be fully divulged to the public at large.

[Narration Two]

Now here you must follow the people in front of you down some
    more stairs

Where as you descend you will find on each side are arranged on
   the walls
More advertisements eager to snatch at your glance as you pass:
If you miss one or two it won't matter, you'll find them again
   further on.

[*Publicity Chorus*]
STRAPLESS BREASTAPPEAL BRA MAKES YOU HARDER TO GET
NEW LYNX LIMOUSINE WITH LOW FAMILY EYELINE
DON'T LET THEM DESCRIBE YOU AS DIRTY! GET 'WET'
HOW'S YOUR COLON LOOK? TREAT IT TO LIQUORICE SOAP
WATCH APPROACH OF PHENOMENAL NEW STAR ON SKYLINE
VAN WORMWOOD EXCLUSIVELY FEATURED IN 'DOPE'
'THIS SOULTWISTER BLISTERS THE PAINT OFF THE SET!'
DRINK MORE DRINK! WEAR MORE CLOTHES! DON'T LOSE HOPE!
   DON'T FORGET!
WEAR MORE SMILES PLEASE! LAUGH LOUDER! LOOK AFTER YOURSELF!
USE CHARM AND DISCRETION! BE TOUGH! DON'T GET LAID ON THE
   SHELF!

[*Train-Wheels Chorus*]
I couldn't care less I couldn't care less I couldn't care less I couldn't
A chance you can't afford to lose a chance you can't afford to lose a
Smooth as glass and tough as hell as smooth as glass and tough as hell
The damned are the damned are the damned are the damned are the
The World to come the Atom Plan the World of Man the Atom Bomb the
Coming Day the Biggest Bang the Wrath of God the Atom Age the Day of
   Wrath . . . (ad inf.)

[*Narration One*]
The Sleeper came here on a Quest, to find that he is lost,
Deepsunk in the confusions of a City underground,
And now looks round him, lonely and bewildered, in the midst
Of anonymous masked multitudes, surrounded by the sounds
Of Latter Pandemonium, Hell's ideal up-to-date
Metropolis of Commerce-cum-Cacophomonium,
The Capital of Every Pseudo Super-City State.

[*Commentator*]
Tonight is Carnival Time in this great underworld city of platforms and
staircases and here I am on the spot to give you a ringside description of the
scene in the Pluto Plaza, where a vast number of masked revellers are already
waiting on the great black ice ballroom floor for the New Season to be officially

*declared open by—why yes, here he is, it's a top secret but I think I can let you
in on it, it's a very important V.I.P. indeed, now I can see his flaming
whiskers and gaily pointed tail as he goes past on his way to the rostrum.
Everyone's tense with excitement, the ice of the ballroom floor's going to melt
in a moment, I think he's going to address them, yes, now here it comes, this is
the moment everyone's been waiting for, you're actually going to hear the Old
Man himself speaking.*

[*V.I.P.*]

*I have every hope that those of you who hear me speak tonight will be as
deeply stirred as I have been to learn that it is to be my special privilege to
have the honour of presenting to Charity for auction on your behalf this most
artfully designed and purposeful-looking Pair of Silver Ceremonial Scissors,
having first severed with them in a single snip—the mile-long cordon-bleu
communication-ribbon which has been arranged so as to run round these
entire fully licensed premises.*

(He cuts the cordon)

*I hereby declare endless Carnival to be left open to the Four Winds of
Publicity, Gossip, Idletalk, and Rumour, and have much sly pleasure in
handing over all responsibility for the conduct of further proceedings to the
Master of Spring Opening Ceremonies, who is already seizing the
Microphone to Address you.*

(Applause)

[*Master of Spring Opening Ceremonies*]

Applause comes first! That's what I like to hear! Just one more
    burst! Now when
I give the sign, let there be music. Bandsmen may burst their drums
    but have no fear,
Dear Dressdesignstars and neat Grooms. Dance, dance until you
    faint.
Abandon everything. No one would think that *your* death might be
    near.
Have no anxiety at all. You'd look a million dollars at your worst.
Never let laughter falter lest its note sound forced, nor let your feet
Trip the less lightly over foolish fear; no one looks quaint
By being opulently over-lightly clad. Dance in the street!
Let the rare joy of true extravagance in dress carry you on
From whirl to whirl, and through hall after hall
Of topflight fashion, as from square to square dance floor!

May I remind you that there are none so mad
Among these streetwalkers that the red carpets spread
For your fleet crystal-slippered toes alone to tread
Will not inspire in them a rapt respect while you are revelling; not
    one
Who following your least step close as facsimile permit
Will not wish that she might be at once flash-photo'd dead
Were she but gowned with the unerring taste shown in your very
    shroud!
So fling yourselves headlong into our Carnival, and let your joy in it
Be long as night, and very, very loud!

    [*Chorus of Masks*] (*confusedly*)

Out of this world. Marvellous! Of course, this is sheer Heaven!
    Out out of this World World. Exquisite.
Divine! Out of this World. Heaven!
    Out of this World. Darling! Such heaven!
I simply worship him. Ah, what Heaven! Worship her worship it
    Simply Divine! I do adore to dance!
    Divine! Out of this World! Sheer heaven, my dear, but too divine!
    This world is heaven! Divine! I adore it, Darling!
You do look heavenly! Adorable! I think your make-up's too divine!

    [*Narration One*]

Although the style's incongruous, one may quote here, I hope,
These apposite Augustan lines from Alexander Pope:

    'Hell rises, Heaven descends, and dance on earth:
    Gods, imps, and monsters, music, rage and mirth,
    A fire, a jig, a battle, and a ball;
    Till one wide conflagration swallows all.'

    [*Voice of a Mask*]

Smoothburnt by artificial sunrays, cold with sweat
Under our swathed robes' sheaths since zero lies within,
Perplexed apparently by our perdition, inwardly
Rehearsing rigmaroles of self-defensive calumny, we go
The tortuous easy way towards uncertainty out of
The pit of ages past. Ours is harsh music. Masks
Like snailshells are become, the glossy whorled
Concealment we excrete to screen our softness from ourselves.
Should silence fall, we'd shake like withered leaves and surely tell
How easy paralytic souls a prey to terror fall

204

Stonedeafened by midwinter's blasts at last! So endless noise
We need to stuff our burning ears with, huge uproars
Must keep on breaking out lest we should judge
Unwillingly how far and near are all one to the void
Whose dungeon swallows up the instant after our least sound.
When buffeted by pangs of dread of failure, we at once
Wrap blankets of cacophony about us, plucking strings
Of strident resonance to death with frantic fingers, while alas,
The only ground-note to all songs is like the throbbing sob
Of childhood by our cold sophistication throttled, choked
Back in our lying throats, to underlie, pent in our breasts,
Each cry during the long spell of our carnival expelled
To swell the roar that rises with each climax repostponed.

(*The Music, in which the* Dies Irae *has been distinguishable, played
simultaneously with* Boys and Girls Come out to Play, *here reaches the
summit of its crescendo with a high, piercing trumpet note.*)

[*Narration One*]

Sleepers, Awake! Awake from Sleep! Back from the world of
    Shades!
The trumpet sounds, the curtain falls, the fabric strange dissolves
And the familiar scene shows through: the darkened stage
Which is the sleeper's bedroom; the familiar properties
Of daily use arranged around the bed. The ordinary street
Outside the window and its streetlamps in the ordinary night.
You awaken from the Pandemonium of your dream, the midnight
    carnival,
And find yourself in the Dark City of the present day again.

[*Narration Two*]

We think at night. We break the spell of every-day if thought can
    wake
From the deep twilight sleep of thinking darkness light.

[*Narration Three*]

It has been said that in the Marketplace, man sleeps his deepest
    sleep.

[*Narration Two*]

Purely material reality, if reality it were, would be lived in by no
    more
Than animated corpses, dead-alive, with ghosts of thoughts

Haunting their brainpans' coils of cells in an irrational way,
However rational their words and meanings were.

[*Narration One*]
Tonight you in the dark attentive to the Night
Thoughts we have here assembled, may be more
Than merely thinking that you wake. When the new day
Emerges from the everlasting East perhaps you may.

3 ENCOUNTER WITH SILENCE

[*Narration One*]
Night Thoughts. Night Music. Now from buried labyrinths and caves
of the town-dweller's anxious dream, from claustrophobic corridors of
nocturnal soliloquy, we move away until we can emerge into the open
air in a secluded countryside.

[*Narration Two*]
There we shall find again the calm night world of Nature.

[*Narration One*]
Nature, the Earth, Unconsciousness and Death. We are drawn down
and back towards them in the Night.

[*Narration Three*]
Nocturnal Music. Meditations in dark gardens. Gradually forming
thoughts pursued in gardens by such solitary strollers as may now find
themselves outdoors, taking a turn or two before retiring, taking a
breath or two of fresher air.

[*Narration One*]
Walking there without a predetermined object; in the starlight; at a
slow pace, uncertainly. Standing still from time to time as though to
listen, yet not listening to any clearly determined sound.

[*Narration Two*]
The Night music has drifted off into remote serenity, leaving the
hearer standing still to listen to the stillness of the garden, waiting to
hear what may be born out of the stillness.

[*Narration Three*]
He stands still and seems to listen to some unknown distant thing;

something that might be coming from . . . from where? What echo from beyond what last horizon?

[*Narration One*]

There is nothing to be heard. The garden is quite still. There is only silence in the darkness.

[*Narration Two*]

There is seldom experienced anywhere on the inhabited earth, for more than a moment or two at a time, such a thing as silence. For it is something we imagine only, Silence, an idea we have of what a complete absence of sound would be like. Real Silence is the message spoken to us that we fear most of all to hear. What we usually call silence is most often no more really than a confused medley of diminutive sounds to which it would be too tiring to pay conscious attention.

[*Narration Three*]

Everywhere about us, day and night, goes on the eddying stream of murmur: little drifting sighs and rumblings, whispers, coughing, whistles, moans. Goes on rising from the earth, the home of life, birthplace of restlessness, where all the rhythms meet, and cross, and intertwine uninterruptedly.

Chorus 1: A window rattling in the wind
Chorus 2: That everlasting rear-exhausting, gear-exhausted car
Chorus 3: Bark of a mongrel
Chorus 1: Tap of an old benighted blind-man's cane
Chorus 2: Another mongrel's barking
Chorus 1: An infinitesimal insect's lovesong, scarcely a second long
Chorus 2: That wretched child . . .
Chorus 3: An ancient iron engine shunts and shunts
Chorus 1: O the wind and the rain in the rain and the wind in the
            rain in the wind
Chorus 2: O love return, return, O darling come . . .
Chorus 3: A mammoth feather's smothered fluttering
Chorus 1: And screams like hell and shunts and shunts and shunts
Chorus 2: Bark of another mongrel
Chorus 3: The same everlasting car

Chorus 1: Old oak's slow taut-slack creak, clock's low quick-slow-
quick tick
Chorus 2: Sand trickling underneath the door, dust blown across
the floor
Chorus 3: The sleeper's snore soon swells the stream which never
dies away
But flows on till with dawn it joins the streaming sounds
of day.

[*Narration One*]

Night music of mysterious hazard. Dream-fugues: variations on
fortuitous themes; intricate tracery unwinding like designs drawn in a
trance across the taut sky of the universal Ear.

[*Narration Two*]

Decrepid gust-blown tinkling of a crumbling pagoda's bells . . .

[*Narration Three*]

Intensely complex tight-screwedup tattoo of tiny drums . . .

[*Narration One*]

The velvet-padded hammering of life-blood's changing pulse.

[*Narration Three*]

The pulse of changing life is the deep underlying constant. And the
Unchanging also is a pulse, flowing through all that lives, a single
pulse.

[*Narration Two*]

The changes and the pauses and occasional recurrence of abrupt
irregularity make sound-patterns we overhear but never really hear.
Our hearing intercepts no more than one bar at a time. These patterns
are upon a scale not measurable in hours. Attention wanders; thinking
intervenes.

[*Narration One*]

The boundaries of the senses are not often clearly realized. The Infra
and the Ultra are fields easily forgotten. Out of hearing stays
unthought-of; out of sight is out of mind. And yet, how haunted we all
are.

[*Narration Two*]

The nightwalker, on a terrace in the garden, unaccompanied, hardly aware of it, half hopes to overhear—that haunting thing. Something that hovers, maybe hovers only just beyond the rim. A thing he has not thought of yet, that no one ever heard.

[*Chorus 1*]

The weir, the misty distant falling waters of the weir among the meadows, make a whispering that swells and faints but never quite subsides.

[*Chorus 2*]

The City blazing with electricity just over the horizon flings its glare-reflection like a continual exclamation of astonishment into the sky, emitting intermittently a high-pitched filtered rumour of its roar.

[*Chorus 3*]

The whisper drifts, the faint roar flutters in the upper air. Both rise and fall. And presently a sudden fine and quite unearthly whistling sound comes sliding down from emptiness, lasting no longer than it takes a shot star's dust to drift and disappear.

[*Chorus 1*]

And then a brisk salt wind blows from the other side of the black downs, and for a while the sea in its perpetual passion of frustration at the shore is to be heard vociferating.

[*Chorus 2*]

A salt breeze seems at last to bring some echo of that sound.

[*Chorus 3*]

Of ocean's ebb and everlasting obstinate resurgence, from afar.

[*Narration One*]

On the terrace in the garden, the solitary stroller has at last come to a standstill. He leans over a parapet and gazes out ahead into the starlit tranquil dark. He thinks of nothing. He lifts his head and gazes and is blind. His heart beating strikes midnight. He breathes in the night's ancientness and freshness, slowly absorbing strength and courage for a coming time when he will have to be reborn.

I stand here staring into darkness and see nothing. Yet it is not nothing that stretches before me away there for ever in whatever direction I turn my eyes. It is the Universe. It is I myself that am nothing. Through my eyes, nothing gazes at Reality, that utterly unqualifiable Something. And slowly the question rises out of nothing's depths, Can I be real if I remain unseen? If I speak out of my innermost reality, shall I not be heard? Why should it be more extraordinary that I who am nothing may be none the less perceived, or that my speaking may be heard, than that nothingness should wonder, gaze and listen?

I stand here speaking of my nothingness; and yet I am a man. It is my heart that speaks, abasing itself in dread before that colossal inscrutability; overwhelmed by the total evidence that what is there must be. I cannot understand however I am able to address what faces me, and yet I know I somehow must respond. From out of that profound night-blue abyss of starry vacancy comes the command: 'Lift up your heart! . . .' I raise my spellbound head and face to face with what I cannot name I worship and adore. I lift my heart up and it speaks my prayer.

O Being, be! O be what faces me, to whom my heart may speak.

Almightiness, O be the Face that bent over me, O be aware and hear.

Acknowledge me, accept me, and may my response responded to help me slowly to realize how we are thus akin.

O be the One, that I may never be alone in knowing that I am. Let my lost loneliness be illusory. Allow to me a part in Being, that I may thus be part of One and All.

I am a man of a benighted century, famished for light and praying out of darkness in the dark. I do not really any longer know what praying means. To pray by rote, repeating time-deconsecrated words, seems vanity to me. I cannot bear to hear myself repeating words of prayer that might be mumbled and not meant. Men of this time seem not to know that there is meaning, or that Being is. All of us talk and talk of all and everything, and shut ourselves up in ourselves and with the curtain of our words shut out the fact that we are blind and dumb. We are afraid of silence, and afraid to look each other in the eye. Talking, we do not speak to one another; one who speaks of many others, seldom fails to disparage them all indiscriminately. Many speeches are made to urge us on to secure peace through understanding; but I will speak no more of speaking: Man has become above all the most indefatigable mimic of all the ways of being man that have ever been thought striking. Men imitate, and I am imitating them. I say 'Man' and 'men' and thus invest abstractions with all my own deficiencies and think I somehow thus may be absolved of the whole failure to be truly man. I am a man. I cry out of my darkness. I could not cry if I were in complete despair.

[*First Voice*]

In the gardens of the Night, breathed on by newly freshened air, wrapped in the sheltering arms of shadows cast by slowly growing things, the consolation of profound Serenity is to be found. Here, in forgetting by degrees the crude immediacies of day, talk's trivialities, the well-worn props and tokens of habitual routine, it is possible to recall to mind and to draw near again to something vastly fundamental, self-effacingly withdrawn, that has been lying there and is there all the time. It is an ever-new discovery to find it still awaiting our return, unsmiling, taciturn, yet limitlessly tolerant and all-comprehending, ready to take us back into obscurity, to share with us its poverty, to close and soothe our eyes.

[*Second Voice*]

The Earth, Nature, Unconsciousness and Death. We are drawn down and back towards them in the Night. But there is Vigil where the walker in the gardens stands and wonders in the dark.

[*First Voice*]

Now the man who spoke aloud just now out of his dark into the darkness: (to no one? to someone? the mystery is not mine to solve that each must face alone) the man who had said: 'I could not cry if I were in despair', turns presently towards the lighted windows he had left behind him earlier, and slowly makes his way back through the scented plants and dangling leaves of the dumbly sleeping garden to his wife and home, his books and bed.

[*Second Voice*]

And as he goes, begins to realize that something has changed in him. The open air, the space about him had first stirred his heart, he lifted up his heart and it had opened, and the wind that blows when it will and comes from nowhere that we know and passes on as unaccountably, had inspired it with its own more vital, lighter, unrestricted and revivifying breath. Silence had delivered its essential message to him, and he had responded. Now he feels that he no longer has the need to reassure himself with words.

[*Third Voice*]

He goes back to his house, he returns to his wife and children. The children have long been asleep upstairs. His wife is sitting where he left her, under the reading-lamp. She closes her book as he enters, looks up at her husband and smiles slowly at him, sleepily. He kisses her.

[*First Voice*]

They are together. The primary division of the human family at night is that which sets those who are alone apart from those who are together. And yet all are alone, as the man realized earlier in the garden; and all those who are isolated in their solitude are really alone only because they do not actually realize the presence of other beings like themselves in the world.

[*Second Voice*]

Greetings to the solitary. Friends, fellow beings, you are not strangers to us. We are closer to one another than we realize. Let us remember one another at night, even though we do not know each other's names.

X
LATER POEMS
(1958–1986)

HALF-AN-HOUR
To Meraud Guevara

. . . and grass grows round the door. The ground,
Without, is grained with root and stone
And yellow-stained where sunlight pours on sand
Through listlessly stirred chestnut-leaves.
This is the long-sought still retreat,
This is the house, the quiet land,
My spirit craves.

   A burning sound,
Uninterrupted as the flow of high-noon's light
Down on the trees from whence it emanates,
The song of the cigales, slowly dissolves
All other thought than that of absolute
Consent, even to anxious transience.

AIX-EN-PROVENCE

REMEMBERING THE DEAD

In the mornings, the day-labourers must set to work once more, and daily tasks be newly undertaken or resumed; and they who work must disregard their usual disillusionment.

'We shall not see a culmination of these labours; our handiwork will not last long nor our success outlive us; our successors taking over what we've done will as like as not disparage it; and if we build houses, they are for strangers to live in for a little while or for the next War to destroy.

'Meanwhile we lose ourselves with a will in what we do today. We tacitly discourage those who would recall too many things or pay too much attention to the future. (All that we cannot see is very small and unimportant). We will put guilt upon them and they shall be silenced.'

And in the mornings, nevertheless, in such a year as this when rain has early in the season put an end to all hope of another extravagant Summer (since a year or two ago an unexpectedly Elysian climate did for once tranform the country with such profusion and intensity of flower-hues and foliage that for the first time many millions were amazed by earth's magnificence); on wet summer mornings, when electric light has to be turned on in the offices in the City, and listlessness and resignation walk the streets, some of the workers (no one knows how many but they may be very numerous) are disturbed by thoughts they have not thought themselves, distracted at their work as though by voices from beneath the chilly ground.

Think, ah! think how vastly they outnumber us by now, the populations of the underworld! How immemorially have they been accumulating there, and how enormous must their number be whom there are none now living to remember. Think how they too may all be working.

I think they think of us—Oh, how incalculably much more than ever we think of them! We scarcely think of them at all; we all prefer soon to forget; if we remember, it is only with regret. They think of us, they think of all of us; they think critically, no doubt, perhaps constructively, with more understanding than we have. Perhaps all day, all night, uninterruptedly.

It may be that only they fully realize that there is no other way of solving the problems of life and death than by thinking about them always.

We do not know the whole Truth; we think we know the Truth. We cannot know it, yet we must. We must seek the Truth we do not know, nor can know while we are still searchers here. Those who have neither curiosity nor doubts are the only real dead.

ON REREADING JACOB BOEHME'S 'AURORA'

Now no one can deny
That what the blessèd shoemaker foretold
Is come about indeed. Babel stands builded high
About us. Nothing avails to save
The old world like a brand from burning. We must die
Before our eyes can see. The dead must live
Before lament and mourning cease to be
The only song heard rise from earth's vast grave.

All shall at last affirm
The Being Boehme faithfully recalled
To have become again real at the final term
Of chaos. Out of the triple void
Of no religion, no communion, no hope, Boehme
Foresaw the sun at midnight would be seen
To rise with rays like healing wings and shine
On the whole world man's fears had else destroyed.

THE GRASS IN THE WASTE PLACES
To Danilo Dolci

What does the grass say?
The Buddha's smile will never tell us quite.

No propaganda, no 'ideas'.

Grass, grasses, fields, the field, 'la terre', our home.
All flesh,
    'cut down, dried up, withereth . . .'

Teeming, brave, swayed by the wind,
Sweet in the shine and shade.

Grass and flowers. Weeds and tares.

Anarchy the law of nature.

A blade of grass glistens with dew
That the Franciscan sun devours.

UNFINISHED POEM FROM ELSEWHERE

Our gentle sister memory,
Our brother Brute Desire,
Conspire from time to time in Time
To set the World afire.

They never can destroy this world,
The strange incestuous Pair;
Their loving intercourse takes place
Within their Father's care.

Their Father is a whoring man,
Although his hair is grey.
He will not set his scythe aside
Till he has had his play.

He loves his fornicating Twins,
He loves to watch them sin;
But hammers at the window-pane
And breaks out once he's in:

'How dare you break my sacred laws!
How many times a day
Have I had to repeat to you
What all the Great Books say:

'There is a limit to all things:
Ye cannot but comply.
What I have told you once for all
Ye shall believe or die.

'You shall not suffer in your bed,
But only in the Bath
Which I am going to prepare
As for an aftermath.

'This afternoon, before the Sun
Has trod his wearying way
Toward his final resting-place,
You shall resume your play.

'But first I'll strip you to the bone
To see in what clean dew
You have been washing in like snow
No spring rains can renew.

'There *is* a mighty Mystery,
In spite of your belief
That Time and old Eternity
Were nothing but a grief.

'The griefs that come, the joys that go,
The vague material things,
Are nothing to the forms of me
Our greater Future brings.

'So get you ready for the Bath
Of Blood you're going to take,
Since, as I happened to pass by,
I saw your Great Mistake.'

WHALES AND DOLPHINS
A Poem for the Greenpeace Foundation

We're told that we must never anthropomorphize
When we are writing about animals, or 'creatures'
as we'd prefer to say; nor are we now allowed,
of course, to speak of 'all God's creatures' either,
since there are few today who can believe
that He exists and once created them and us.
To write a poem about whales or dolphins, then,
presents a challenge to all those who see
in the great whale the dread Leviathan
which Scripture teaches man should look upon
as the huge proof of the Creator's mightiness,
the ruler of the deeps, and in the guise
of the White Whale of Melville's 'Moby Dick'
a mighty symbol of both Death and Mystery;
or who, as I do, see in the dolphin's face
the look both of the cherubim and of the unborn child
safe in its mother's womb, with the angelic, innocent
smile worn by all the creatures of God's Paradise.
Many the myths about the dolphin. *Dauphin* means,
or used to, dolphin and also 'first-born'; and
a boy upon a dolphin's back is such an old
image, it surely tells us men have always sensed
some sort of kinship that the reason can't explain
between the amphibious being and our own.
Then there's the recent question or new myth
about the dolphin's sort of speech: a mystery
indeed! Poets and thinkers are increasingly
concerned with the great problems language sets.
A poem should avoid abstraction and all forms

of private declaration of belief; yet I must state
that I'm convinced by what is called the Fall of Man.
We've been turned out of Paradise; we've made the world
into a shambles and a slaughter-house; we've lost
the primal *Urspräch* which may once have been
also an aid in our communion with the beasts
we now exploit and prey upon. Polluted earth,
polluted souls: Now finally, perhaps too late,
we try to care, if not to pray, for some Salvation.
A poet friend of mine[1] wrote lately that: 'We live
in the mind of God, here, now and always, for there is
no other place.' And R. Buckminster Fuller wrote
in Nineteen Sixty-three: 'Stop "calling names"
names that are meaningless; you can't suppress God
by killing off people which are, physically,
only trans-ceiver mechanisms through which God
is broadcasting.' And too: 'The more man becomes man,
the more it will be needful for him to,
and to know how to, worship': thus the Père
Teilhard de Chardin. I do not digress.
If you have faith you may not have it every day
but somehow you believe that we shall not destroy
ourselves and God's creation; though we can
'kill off people' and, be it added, species like
the direly menaced whales and dwindling dolphins.
Now 'the light of the public darkens everything.'[2]
But still the animal kingdom and the world of nature can
remind us of our long-lost innocence. All things shall be
made new. Let chaos come. The mortal must first die.
Yet even an atheist poet[3] could write: 'The rose
tells that the aptitude to be regenerated has
no limit': and, 'what selectivity there can occur,
only just in time, and succeed in imposing its law
in spite of everything. Man sees this pinion tremble
which in every language is the first great letter of
the word Resurrection.' Redemption. Paradise Regained.
God's Kingdom here on earth. Absurd, discarded dreams?
Not only fools can still believe and fight for faith
and meaning: to preserve our innate, obstinate capacity
for love, for wonder at the miracle of life:

[1] Kathleen Raine
[2] Martin Heidegger
[3] André Breton

to speak out even if the words one's forced to use
seem worn nearly to death, and say: Yes, we can still
do what we can to preserve not only such rare things
as whales and dolphins, but the eternal Mystery of which
they are both emblem and incarnate form.

PRELUDE TO A NEW FIN-DE-SIÈCLE

Incessant urging, curt, peremptory:
Write what you will, in verse or otherwise,
Intelligible, using simple metaphors.
Address a reader not just hypothetical
But flesh and blood in no need of harangues.
The time has come. We're on the very brink
Of what? Can any prophet, true or false,
Make himself heard above the mad uproar
Of all the mingling and ambiguous,
Self-righteous or dismayed denunciations,
Warnings and dire predictions that assail us from
All 'informed sources', media-debased and bent?

—If this is a poem, where are the images?
—What images suffice? Corpses and carrion,
Ubiquitous bloodshed, bigger, more beastly bombs,
Stockpiled atomic warheads, stanchless wounds,
Ruins and rubble, manic messiahs and mobs.
—But poets make beauty out of ghastliness . . .
—You think I want to? Think truth beautiful?
—'A terrible beauty is born . . .'—It is indeed.

In youth I did in spite of everything
Believe with Keats and Shelley such things as
That poets can 'legislate' and prophesy;
Or like Stravinsky when he wrote 'The Rite'
Become transmitting vessels for new sounds
From an inspiring, unknown world within.
I'm over sixty now, my dubious gift has gone,
I can but grope for unexpected similes.

But now as in the 'Thirties I can once again
Feel passion and frustration and that sense
Of expectation, imminence and pressing need
To express something that just must be said.
Mature awareness knows that poetry
Today demands the essence and the minimum;
That only Silence such as God's could say the Whole.
One stark vocabulary at least remains.
The litany of lurid headline-names
Merely to mention which can nag the nerves:
Vietnam, Angola, Thailand and Pakistan,
Chile, Cambodia, Iran, Afghanistan,
Derry's Bogside, Belfast and Crossmaglen;
Up in Strathclyde or down on Porton Down,
On Three Mile Island or in Seveso Italy
Then there are Manson, Pol Pot and Amin,
To name at random just three myth-monsters,
Too many more to mention, all mass-murderers:—
None of them need an adjective and though we're sick
Of being sickened by them they will stay engraved
Or branded on even callous consciences.

And yet I yearn to end by trying to evoke
A summer dawn I saw when I was not yet eight,
And having risen early watched for an hour or more
A transcendental transformation of auroral clouds,
Like a prophetic vision granted from on high.
I cannot see much now. The dawn is always new
As nature is, however much we blind ourselves and try
To poison the Earth-Mother. But an ancient text
Tells of what I believe may happen soon today:
The raven disappears as night draws to its close,
Then as the day approaches the bird flies without wings;
It vomits forth the rainbow and its body becomes red,
And on its back a condensation of pure water forms.
For that which is above is still as that which is below
For the perfecting of the One Thing, which is now
As it shall ever be, World without End, D.V.

VARIATIONS ON A PHRASE

'le lièvre fit sa prière à l'arc-en-ciel à travers la toile de l'araignée...'
RIMBAUD

The hare sent up his prayer to the rainbow
Through the spider's fine-spun filmy web,
Despite the huntsmen tracking it below.

The hunters set their snares, the norns weave threads,
Hephaestus' net awaits all peccant pairs.
A filament of light through heaven spreads.

A shaft of sunshine transpierces the dust
That rises as the shell's target explodes,
And glorifies it. Deep in mud we must

Unseal our eyes through choking smoke to see
How slaughter and compassion can combine
To trace a liberating filigree.

A hostage prisoned in a stinking cell
For just an instant saw a glinting fly
Above him as a sign from heaven not hell.

In chthonic labyrinth where we now stray
Do Thou in us make peace, O lightbringer.
Submerged in darkness glows the serene day.

While raw-scabbed refugees without end file
Past numbed spectators, an aeon elsewhere
Some insane sanity sustains its smile.

Yet jackals howl across the wastes of thyme.
The drunken boat speeds on. The skilled music
Still needed by desire runs out of time.

The Charleville boy ended up peddling guns
In Ethiopia, amnesic of dream.
We can end roasted by our man-made suns.

RARE OCCASIONAL POEM

May 13th 1982

The 'Thought for Today' that was broadcast this morning
Told us that Crisis means Judgement. But who is the Judge?
You may or you may not believe that one exists.
Judgement can signify verdict, decision or
Fate, among other things. Yesterday, Fatima:
Priest tried to stab Pope. There was one more announcement
That a new Incarnation of Christ will appear
On TV before June has ended; by which time
Perhaps the dense fog which just now envelops us
May have somewhat dispersed, thus revealing at least
Whether fervour for fatherland, freedom or force
Have prevailed in the South Atlantic . . . or foresight.

DODECATRIBUTE TO
MIRON GRINDEA AT 75

Many years, many memories, my dear Miron. . .
I met you early, an ignescent incomer,
Raw yet ready to recognize your rare repute
Of openness to all original output.
Now none can ignore your initiative nous.

Great is our gratitude for your genial gift:
Rampart of rance amidst Ragnarok's rioting,
Indispensable international index,
Nonesuch never needless of normative notions,
Doyen of discerningly diglot dossiers,
Exemplarily edited for an era—
ADAM, acme of annals of authentic art.

A SARUM SESTINA
To Satish Kumar

Schooldays were centred round the tallest spire
In England, whose chime-pealing ruled our lives,
Spent in the confines of a leafy Close:
Chimes that controlled the hours we spent in singing,
Entered the classrooms to restrict our lessons
And punctuated the half-times of games.

The gravel courtyard where we played rough games
During the early break or after singing
In the Cathedral circled by the Close
And dominated by its soaring spire
Saw many minor dramas of our lives.
Such playgrounds predetermine later lessons.

Daily dividing services, meals, lessons,
Musical time resounded through the Close,
Metered existence like the rules of games.
What single cord connects most schoolboys' lives?
Not many consist first of stints of singing.
Our choral rearing paralleled a spire.

Reaching fourteen within sight of that spire
Unconsciously defined our growing lives,
As music's discipline informed our lessons.
We grew aware of how all round the Close
Households were run on lines that like our singing
Were regulated as communal games.

We sensed the serious need for fun and games,
What funny folk can populate a Close.
We relished festive meals as we did singing.
Beauty of buildings balanced boring lessons.
We looked relieved at times up at the spire
Balanced serene above parochial lives.

Grubby and trivial though our schoolboy lives
Were as all are, we found in singing
That liberation and delight result from lessons.
Under the ageless aegis of the spire
Seasonal feasts were ever-renewed games.
Box-hedges, limes and lawns line Sarum Close.

Choristers in that Close lead lucky lives.
They are taught by a spire and learn through singing
That hard lessons can be enjoyed like games.

THALASSA: THE UNSPEAKABLE SEA
For Mimmo Morina

Sitting on a beach facing the foaming collapse of the waves of a vast expanse of acrid water stretching away as far as the distant line that indicates the curvature of the globe

Sitting in a deckchair with ballpoint and notepad facing the theme of Thalassa

Vociferations uninterrupted since the first emergence of all animal life. Thunders—murmurs: furies—calms. Ultimate challenge to language. Total proscription of words

Primal matrix: insatiable grave. Unalterably other. Unlikeness extending out of sight

We are a minority inhabiting an environment unaware of having given us birth

Swimming, sailing and fishing: ephemeral superfluities

How long before the final drowning of that book wherein it is written that our finest order is no more than a heap of garbage dumped at random on the verge of the purest and most polluted of waters, undrinkable and deadly to all but the Kraken and its countless amphibious hordes?

Triumphant rise, fall and crash of a last billow against the definitive deserted shore: all too human imagining that no incarnate consciousness can ever realize.

ENTRANCE TO A LANE

on a painting by Graham Sutherland

To Elizabeth Jennings

Memento rectangled to lead the gaze
From outer levels to a hub of white,
An elsewhere that recedes from coiling planes

Sequestered rural scene reputed Welsh,
Season's regalia reduced to tones
Of veld and verdure, leaves to sprays of blotch

A static vortex wherein ochre glows
Softly in strata linked by streaks and zones
Of compact shade and layers of virid light

The felt-floored lane leads to a blank where hues
And perceptions vanish as fast as time
Into the *non-lieu* beyond mortal reach

Where red is not an opposite of rot
Or devastation the reverse of peace
And all those things that were the case resume.

A FURTHER FRONTIER

Viewed from Corfu

To Lawrence Durrell

Seen across leagues of amethystine calm,
Two facing foreheads, one afforested,
The other sparsely greened as with Greek-hay,
An isthmus vista in between them hazed
By distant fluorescent shimmering
Of drowsy blended colours in which soot
Suffuses violet, peach and ivory.
Far to the East, a tranquil smoulder veils
Some remote city old as Trebizond,
Sated with myth and stunned by history,
Where linger shades of despots, peasants, saints,
Lost in oblivion's drifting dust. The end
Of afternoon approaches, the tenth month

227

Is almost here, further to obumbrate
A land once white with dawn, the nearby shore
Of North Helladic rock, whose dwellers owe
Fealty alike to thoughts of men long dead.
Night hovers like the question haunting all
As to whether *eschatos* has not come:
Unseen above hangs Saturn's fractured scythe.

(1985)

NOVEMBER IN DEVON

Leaving Plymouth last seen after first smashed by bombs,
   Driving North all the morning after rain
      Towards Hartland's hospitable hearth
   Through landscapes clad in disruptive pattern
Material edged by hedge or walls of dry-stone:

Under a cover of commingling cloud and clear,
   Drifts of drab haze transpierced by wet blue slate,
      Between lofty moor and deep glen
   Past lanes twisting off into the arcane
We spin towards midday's strengthening sun.

After Launceston eleven o'clock approaches
   At a thousand revs per minute four times
      Beneath us: the car radio
   Picks up brass playing *Nimrod* in Whitehall,
Rearousing a reticent love for this land.

While memory brings back like a sepia still
   Holding my mother's hand in a Bournemouth
      Doorway during the first of all
   Remembrance Days' two minutes of silence,
Today I anticipate the advent of death.

A parade of folk sporting mass-produced poppies
   In the next village briefly delays us
      At a border-point round which spread
   Areas of age-old non-violence.
In ivy-dark gardens hang white rags of late rose.

An abrupt paranoia wonders just how sure
  One can be now that no secret convoy
    Was out during last night on roads
  Linking Hinkley Point and Bull Head, that near-
by tin-mines or tumuli hide no lethal hoards.

At half my age this might have worried me more.
  The South country kept my childhood secure.
    Now I know that to Whinny-moor
  Before long I shall come, as one more year
Declines towards departure in deceptive calm.

                                        (1986)

OXFORD POETS

Fleur Adcock

Yehuda Amichai

James Berry

Edward Kamau Brathwaite

Joseph Brodsky

D. J. Enright

Roy Fisher

David Gascoyne

David Harsent

Anthony Hecht

Zbigniew Herbert

Thomas Kinsella

Brad Leithauser

Herbert Lomas

Derek Mahon

Medbh McGuckian

James Merrill

John Montague

Peter Porter

Craig Raine

Tom Rawling

Christopher Reid

Stephen Romer

Carole Satyamurti

Peter Scupham

Penelope Shuttle

Louis Simpson

Anne Stevenson

George Szirtes

Anthony Thwaite

Charles Tomlinson

Chris Wallace-Crabbe

Hugo Williams

also

Basil Bunting

Keith Douglas

Edward Thomas